# Lecture Notes in Electrical Engineering

## Volume 1478

The book series *Lecture Notes in Electrical Engineering* (LNEE) publishes the latest developments in Electrical Engineering—quickly, informally and in high quality. While original research reported in proceedings and monographs has traditionally formed the core of LNEE, we also encourage authors to submit books devoted to supporting student education and professional training in the various fields and applications areas of electrical engineering. The series cover classical and emerging topics concerning:

- Communication Engineering, Information Theory and Networks
- Electronics Engineering and Microelectronics
- Signal, Image and Speech Processing
- Wireless and Mobile Communication
- Circuits and Systems
- Energy Systems, Power Electronics and Electrical Machines
- Electro-optical Engineering
- Instrumentation Engineering
- Avionics Engineering
- Control Systems
- Internet-of-Things and Cybersecurity
- Biomedical Devices, MEMS and NEMS

For general information about this book series, comments or suggestions, please contact leontina.dicecco@springer.com.

To submit a proposal or request further information, please contact the Publishing Editor in your country:

**China**

Jasmine Dou, Editor (jasmine.dou@springer.com)

**India, Japan, Rest of Asia**

Swati Meherishi, Editorial Director (Swati.Meherishi@springer.com)

**Southeast Asia, Australia, New Zealand**

Ramesh Nath Premnath, Editor (ramesh.premnath@springernature.com)

**USA, Canada**

Michael Luby, Senior Editor (michael.luby@springer.com)

**All other Countries**

Leontina Di Cecco, Senior Editor (leontina.dicecco@springer.com)

**** This series is indexed by EI Compendex and Scopus databases. ****

Yongsheng Ma
Editor

# Proceedings of 4th International Conference on Intelligent Systems Design and Engineering Applications

## Proceedings of ISDEA 2025

Springer

*Editor*
Yongsheng Ma
Southern University of Science
and Technology
Shenzhen, Guangdong, China

ISSN 1876-1100 ISSN 1876-1119 (electronic)
Lecture Notes in Electrical Engineering
ISBN 978-981-95-1801-2 ISBN 978-981-95-1802-9 (eBook)
https://doi.org/10.1007/978-981-95-1802-9



This Springer imprint is published by the registered company Springer Nature Singapore Pte Ltd.
The registered company address is: 152 Beach Road, #21-01/04 Gateway East, Singapore 189721, Singapore

# Conference Committees

## Conference Chair

Shiho Kim, Yonsei University, South Korea

## Conference Co-chairs

Yungcheol Byun, Jeju National University, South Korea
Makoto Iwasaki, Nagoya Institute of Technology, Japan
Yongsheng Ma, SUSTech in Shenzhen, China

## Conference Program Chairs

Songkuk Kim, Yonsei University, South Korea
Shuangxi Huang, Tsinghua University, China
Teo Tee Hui, Singapore University of Technology and Design, Singapore

## Conference Program Co-chairs

Yow Kin Choong, University of Regina, Canada
Cheng Siong Chin, Newcastle University, Singapore

## Publicity Chair

Euntai Kim, Yonsei University, South Korea

## Technical Program Committees

Ljiljana Trajkovic, Simon Fraser University, Canada
Herbert Ho-Ching Iu, University of Western Australia, Australia
Viranjay Mohan Srivastava, Birmingham City University, UK
Yew Kee Wong Eric, Hong Kong Chu Hai College, Hong Kong, China
P. Geetha, Mohan Babu University, India
Yeungkwun Kim, Western Illinois University, USA
Ahmed Moumena, Hassiba Benbouali University of Chlef, Algeria
Hasan Kadhem, American University of Bahrain (AUBH), Bahrain
Omer K. Jasim, University of Fallujah, Iraq
Radhakrishna Bhat, Manipal Institute of Technology, India
Janvier Kamanzi, Cape Peninsula University of Technology, South Africa
Nandha Kumar T., The University of Nottingham Malaysia Campus, Malaysia
Yagiz Nalcakan, Yonsei University, South Korea
Ioannis Vardiambasis, Hellenic Mediterranean University, Greece
Wanthanee Prachuabsupakij, King Mongkut's University of Technology North Bangkok, Thailand
Amirhossein Khosravipour, Science and Research Kermanshah Azad University, Iran
Iara Margolis Ribeiro, University of Minho, Portugal
Aidrina Binti Mohamed Sofiadin, International Islamic University, Malaysia
Tan Chi Wee, Tunku Abdul Rahman University of Management and Technology, Malaysia

# Preface

We are honored to introduce this volume, the proceedings of 2025 4th International Conference on Intelligent Systems Design and Engineering Applications (ISDEA 2025), which took place in Seoul, South Korea, during April 19–21, 2025.

ISDEA provides an idea-exchange and discussion platform for the world's engineers and academia to share cutting-edge information, address the hottest issue in intelligent systems design and engineering applications, explore new technologies, exchange and build upon ideas.

This year, ISDEA received submissions from several countries and regions, including Italy, China, South Korea, Saudi Arabia, India, Thailand, Philippines, Turkey, Sri Lanka, Ghana, Bahrain, etc. All submitted papers have been reviewed by international reviewers in the field based on the criteria of originality, significance, quality and clarity. Accepted chapters are grouped into 4 parts, with topics: Computer Network and Information Security, Modern Electronic Systems and Signal Analysis, Intelligent Image Analysis and Detection Technology and Machine Learning Models and High Performance Computing. We really hope that this volume of the conference proceedings will serve as a valuable reference for you.

On behalf of the organizing committee, we would like to express our gratitude to all the delegates and all our committee members and reviewers for their continuous support and contributions. We would also like to acknowledge with special appreciation to our speakers for their distinguished lectures about the latest information on intelligent systems design and engineering applications.

We look forward to reuniting with you next year.

Shenzhen, China

Prof. Yongsheng Ma
Proceedings Editor

# Contents

**Computer Network and Information Security**

**Enhancing Enquiry Response Efficiency: Methods for Article Data Extraction in Manufacturing** ..... 3
Tobias Eberhardt, Felix Fuchs, Walter Schaaf, Alexander Verl, and Marco F. Huber
1 Introduction ..... 4
2 State of the Art ..... 5
3 Analysis of Enquiry Communication of Representative Component and System Manufacturer ..... 5
4 Conception ..... 8
5 Realization ..... 10
6 Evaluation ..... 11
7 Discussion ..... 13
8 Conclusion and Outlook ..... 14
References ..... 15

**Go Small and Similar: A Simple Output Decay Brings Better Performance** ..... 17
Cheng Xuan, Gong Haigang, Wang Xiaomin, and Liu Ming
1 Introduction ..... 17
2 Observations ..... 19
3 Method ..... 19
4 Why Does Output Decay Help? ..... 20
5 Experiments ..... 25
5.1 Output Decay Can Be Universally Applied to Various Architectures ..... 25
5.2 Output Decay Is Equally Effective on Multiple Benchmark Datasets ..... 26
5.3 Output Decay Is Compatible with Various Techniques ..... 26
6 Discussion ..... 27
References ..... 27

**Security Framework Development for Adaptive Phishing Detection Using Hybrid Deep Learning Techniques** ........ 31
Mosiur Rahaman, Nicko Cajes, Brij B. Gupta, and Kwok Tai Chui
1 Introduction ........ 32
2 Related Work ........ 32
3 Proposed Approach ........ 34
3.1 Data Collection and Preprocessing ........ 34
3.2 Proposed Framework and Model Development ........ 35
3.3 Selection of Deep Learning Techniques ........ 37
3.4 Hybridization Strategy ........ 38
3.5 Training the Model ........ 38
3.6 Validation and Testing Model ........ 39
4 Experimental Results and Discussion ........ 39
4.1 Results and Discussion ........ 40
4.2 Accuracy and Precision ........ 41
5 Conclusion and Future Work ........ 41
References ........ 43

**Security Implications of Quantum-Resistant Encryption in Emerging IoT Networks** ........ 47
Eric Blancaflor, Mark Francis James Bona, David Josiah Dychioco, Matthew Jake Inson, and Prince Rayly Reyes
1 Introduction ........ 47
1.1 Background of the Study ........ 47
1.2 Significance of the Study ........ 49
1.3 Research Gap ........ 49
1.4 Objective of the Study ........ 50
1.5 Scope and Delimitation ........ 50
2 Literature Review ........ 51
2.1 Identify IoT Security Vulnerabilities ........ 51
2.2 Access Quantum-Resistant Encryption ........ 54
2.3 Examine Real-World Implementation Challenges ........ 55
2.4 Synthesis ........ 57
3 Conclusion ........ 57
References ........ 58

**Modern Electronic Systems and Signal Analysis**

**FOV-Adaptive Bilateral-Branch Network for Etch Bias Model with SEM Contour** ........ 63
Feng Chaoxin, Cheng Qian, and Jia Ruoran
1 Introduction ........ 64
2 Related Work ........ 67
2.1 Traditional Empirical Etch Model ........ 67
2.2 Machine-Learning (ML) Etch Model ........ 67

2.3 CNN-Based Etch Model ... 69
3 FOV-Adaptive Bilateral-Branch Network Etch Model ... 69
3.1 Loading Effect ... 69
3.2 The Effect of Input Size on CNN Model ... 70
3.3 Different Pattern Density Has Different Optimal FOV ... 72
3.4 Etch Macro-Loading Effect ... 73
3.5 FOV-Adaptive Bilateral-Branch Network ... 74
3.6 FA-BBN Network Work Flow ... 76
4 Experiments Data ... 77
4.1 Technical Details ... 78
4.2 Analysis ... 79
5 Conclusions ... 82
References ... 82

**IoT-Based Smart Agriculture and Crop Recommendation Using ML** ... 85
Simran Jaggi, Sunil K. Singh, Sudhakar Kumar, Anish Sharma, Varsha Arya, Kwok Tai Chui, and Brij B. Gupta
1 Introduction ... 86
2 Literature Review ... 88
3 Challenges and Open Issues ... 89
4 Proposed Methodology ... 89
5 Results ... 94
5.1 Experimentation and Evaluation ... 94
5.2 Comparative Analysis ... 94
6 Conclusion and Future Scope ... 94
References ... 96

**Research on Remaining Life Prediction of the Reactor CRDM Based on Monitoring Signal Reconstruction and Adaboost-CNN** ... 99
Liu Zhi-long, Zeng Zhen, Wang Qin-chao, Huang Yan-ping, Nie Chang-hua, and Xie Tian-zhou
1 Introduction ... 100
2 Introduction Noise Reduction and Correction of CRDM Monitoring Signals ... 101
2.1 Noise Reduction of Monitoring Signals Based on CEEMDAN ... 101
2.2 Correction of Denoised CRDM Signals ... 103
3 Introduction Noise Reduction and Correction of CRDM Monitoring Signals ... 104
3.1 Convolutional Neural Network ... 104
3.2 Adaboost ... 105
3.3 Residual Life Prediction Method and Model Performance Evaluation ... 106
4 Experimental Results and Analysis ... 107
5 Conclusion ... 110
References ... 110

**Model-Based Prognosis for Hybrid Systems: A Perspective of Patent** ... 113
Tsai Hong-Wen
1 Introduction ... 113
1.1 A Subsection Sample ... 113
2 Discussion for the Selected Issued Patents ... 115
2.1 Fault Models ... 116
2.2 Data-Driven Analysis ... 117
2.3 Simulation and Testing ... 118
2.4 System-Level Integration ... 119
2.5 Optimization and Adaptation ... 120
3 Discussion ... 120
References ... 121

**From Preservation to Innovation: A Comprehensive Review of Generative AI in the Design Research of Traditional Heritage Furniture** ... 123
Wang Yali, Xi Yuchen, Liu Xinxiong, Gan Yan, and Xiang Xingrui
1 Introduction ... 124
2 Research Framework and Methodology ... 125
3 Bibliometric Analysis with CiteSpace ... 126
3.1 The Specific Operational Steps of Scientometrics ... 126
3.2 The Specific Operational Steps of Scientometrics ... 127
3.3 Knowledge Map and Keyword Clustering ... 128
3.4 Evolution of Hot Topics and Cutting-Edge Insights ... 133
4 Core Topics and Technical Insights ... 137
4.1 Interaction Between Generative Models and Cultural Heritage ... 137
4.2 Application of Generative Models and Innovative Design ... 139
4.3 Integration of Generative Models and Furniture Design ... 140
5 Results and Discussion ... 142
References ... 145

**Intelligent Image Analysis and Detection Technology**

**Surface Defect Detection Model for Quick-Connect Fittings** ... 149
Lin Zhou, Chen Wang, Shuai Yang, and Peng Huang
1 Introduction ... 150
2 Related Works ... 151
2.1 Enterprise Testing Process ... 151
2.2 Cloud-Edge-End Collaborative Architecture for Inspection of Surface Defects in QCF ... 152
2.3 Data ... 153
3 Modeling ... 154
3.1 C3GNet ... 154
3.2 Neck Model ... 156
3.3 Head Adding a Small Target Detection Layer ... 158

3.4 Unsupervised Learning Model .... 159
4 Experiment Detail .... 160
4.1 Experimental Environment and Parameter Configuration .... 160
4.2 Experiment Platform .... 160
5 Analysis of Datasets .... 161
5.1 Ablation Experiment .... 161
5.2 Fusion Experiment of DySample .... 162
5.3 Contrast Experiment .... 163
6 Conclusion and Discussion .... 167
References .... 167

**Designing an Origami-Inspired Soft-Actuated Smart System for Portable Pelvic Rehabilitation** .... 169
Alessia Buffagni, Elisa Venturini, and Giovanni Borga
1 Introduction .... 170
2 Method .... 171
2.1 Current Tools for Pelvic Rehabilitation .... 171
2.2 Limitations of the Current Tools and Need for Targeted Solutions .... 172
2.3 Advancements in Soft Actuators for Biomedical Applications .... 173
3 Results .... 175
3.1 Carefit, Origami-Inspired Soft-Actuated Smart System .... 175
3.2 Technological Components and Operation .... 178
4 Discussion .... 179
5 Conclusion .... 183
References .... 183

**Top-Forging Crack Detection for Cold-Heading Steel** .... 187
Long Wu, Yu-Xiu Wu, and Hong Wei Yang
1 Introduction .... 187
2 YOLOv8n Improved Model .... 189
2.1 Optimization of CBAM and C2f Modules .... 190
2.2 DAttention Module .... 191
2.3 Loss Function Optimization .... 192
3 Experimentation and Analysis .... 193
3.1 Performance Indicators and Ablation Experiments .... 193
3.2 Algorithm Comparison Test .... 194
3.3 Confidence Level Comparison Test .... 196
4 Conclusion .... 196
References .... 197

**Revised Local Laplacian Filter for Turbid Underwater Image Enhancement** .... 199
Shunsuke Takao and Taketsugu Hirabayashi
1 Introduction .... 200
2 Related Work .... 202

2.1 Underwater Image Enhancement Methods ..................... 202
2.2 Geometry Reconstruction of 3D Underwater Structures .......... 202
3 Method ........................................................ 203
3.1 Local Laplacian Filter (LLF) ................................ 203
3.2 Discrete Wavelet Transform (DWT) ............................ 204
3.3 Scheme of Proposed LLF-Fusion .............................. 204
4 Numerical Experiments ......................................... 205
4.1 Settings of Numerical Experiments ........................... 205
4.2 Comparison to Other UIE Methods ............................. 206
4.3 Insights on LLF-Fusion and Ablated Results .................. 207
5 Physical Experiments .......................................... 209
5.1 Collection of Hazy Pictures of Color Chart ................. 209
5.2 Underwater Structure from Motion in Turbid Water ............ 210
6 Conclusion .................................................... 211
References ...................................................... 211

**Lightweight UAV-Based Road Disease Detection Algorithm** ............ 213
Guo Jinbo, Wang Shenghuai, Xu Fenghua, Chen Xiaohui, Wang Chen, and Zhang Wei
1 Introduction .................................................. 213
2 YOLOv8 Algorithm .............................................. 215
3 LHP-YOLO Algorithm ............................................ 216
3.1 Lightweight Network Architecture Design ..................... 216
3.2 Knowledge Distillation ...................................... 219
3.3 Dataset and Experimental Setup .............................. 220
3.4 Evaluation Criteria ......................................... 222
3.5 Ablation Experiment ......................................... 222
3.6 Visualization Analysis ...................................... 223
3.7 Contrast Test ............................................... 223
4 Conclusion .................................................... 225
References ...................................................... 225

**Machine Learning Models and High Performance Computing**

**Analysis of Multi-model Fusion Strategies and Precision Optimization in Drug Score Prediction** ............ 229
Ran XiangDong and Zhang Yu
1 Introduction .................................................. 229
1.1 Contributions ............................................... 230
1.2 Related Work ................................................ 230
1.3 Comparative Analysis ........................................ 231
2 Methodology ................................................... 232
2.1 Data Preprocessing .......................................... 232
2.2 Model Architecture .......................................... 233
3 Experiment Setup .............................................. 234
3.1 Hyperparameter Settings ..................................... 234

3.2 Evaluation Metrics ........ 235
4 Training Process ........ 235
4.1 Training and Validation Loss over Epochs ........ 235
4.2 Validation Set Performance ........ 236
4.3 Test Set Performance ........ 236
4.4 Summary ........ 237
5 Ablation Experiments ........ 237
5.1 Experimental Setup ........ 238
5.2 Experimental Results and Analysis ........ 238
6 Discussion ........ 239
References ........ 240

**Optimization of Feature Extraction and Enhancement for High Resolution Data: A Hybrid Approach Combining Classical and Neural Network Technique** ........ 243
Hazel San Loverez-Patilano and Melvin A. Ballera
1 Introduction ........ 243
2 Review Related Literature ........ 244
3 Methodology ........ 245
4 Results and Discussion ........ 246
4.1 Feature Extraction and Enhancement Framework ........ 246
4.2 Binary Large Object (BLOB) Detector ........ 246
4.3 Feature Extraction ........ 247
4.4 Feature Matching ........ 248
5 Experimental Result ........ 248
6 Conclusion ........ 252
7 Recommendations ........ 252
References ........ 253

**A Study of Generative Artificial Intelligence (GenAI) Capabilities in Generating Supervised Machine Learning Models** ........ 255
Yousif Abuzuhaira, Mohamed Badawy, and Hasan Kadhem
1 Introduction ........ 256
2 Related Work ........ 256
2.1 Performance in Programming and Coding ........ 257
2.2 Contextual Adaptation and Evolution of Thought in LLMs ........ 257
2.3 Prompt Engineering Tools ........ 258
2.4 LLM Feature Generation ........ 258
2.5 Performance of LLMs ........ 258
3 Datasets ........ 259
3.1 Regression Dataset: Laptop Price Prediction ........ 259
3.2 Classification Dataset: Portuguese Bank Marketing Outcome Prediction ........ 260
4 Methodology ........ 260
4.1 Phase 1: Training Models Locally ........ 260

4.2 Phase 2: Evaluating LLMs' Capability to Generate Machine Learning Models ........ 262
5 Results and Discussion ........ 264
5.1 Phase 1: Training Models Locally ........ 264
5.2 Phase 2: Evaluation of LLMs ........ 266
References ........ 276

# Computer Network and Information Security

# Enhancing Enquiry Response Efficiency: Methods for Article Data Extraction in Manufacturing

**Tobias Eberhardt, Felix Fuchs, Walter Schaaf, Alexander Verl, and Marco F. Huber**

**Abstract** Efficient and accurate processing tracking of customer enquiries is crucial in terms of customer satisfaction and business success. Answering these enquiries is often done manually and is therefore linked with high personnel costs in the sales department. In a first step, this paper analyses the customer enquiries of a representative component and system manufacturer. In a second step, the paper compares four different concepts for the automatic extraction of relevant data based on real customer enquiries to generate a product recommendation. A hybrid approach consisting of a lexicon, regular expressions and a Large Language Model delivers the best overall results.

**Keywords** Customer enquiries · Article data · Large language model · Lexicon · Regular expression

T. Eberhardt (✉) · F. Fuchs
Graduate School of Excellence Advanced Manufacturing Engineering (GSaME), Stuttgart, Germany
e-mail: Tobias.Eberhardt@gsame.uni-stuttgart.de

T. Eberhardt · A. Verl
Institute for Control Engineering of Machine Tools and Manufacturing Units (ISW), Stuttgart, Germany

F. Fuchs · M. F. Huber
Institute of Industrial Manufacturing and Management (IFF), Stuttgart, Germany

M. F. Huber
Fraunhofer Institute for Manufacturing Engineering and Automation (IPA), Stuttgart, Germany

T. Eberhardt · W. Schaaf
J. Schmalz GmbH, Glatten, Germany

 
Y. Ma (ed.), *Proceedings of 4th International Conference on Intelligent Systems Design and Engineering Applications*, Lecture Notes in Electrical Engineering 1478,
https://doi.org/10.1007/978-981-95-1802-9_1

## 1 Introduction

Nowadays in the manufacturing industry, efficient and accurate handling of customer enquiries is a key factor in customer satisfaction and business success [1].

This study focuses on component and system manufacturers, as they are typically the first to be consulted at the outset of the machine life cycle, and their products often constitute integral components of subsequent machines, such as robotic systems. Figure 1 illustrates this process from the standpoint of a machine manufacturer. Component and system manufacturers face the daily challenge of dealing with many customer enquiries, often with specific questions about items. These enquiries typically include part numbers and part designations, but also important contextual information such as quantity, which must be correctly extracted and processed to ensure a timely and accurate response [2].

Traditionally, these customer enquiries are handled and answered manually by internal sales staff. Due to the large number of daily enquiries and the associated effort, precious company resources are tied up. This paper analyses the effectiveness of different approaches for automatically extracting relevant article data from customer enquiries to enable assisted or automated responses to customer enquiries.

To this end, we first analyze typical customer enquiries and then compare a pure Large Language Model (LLM) approach with an expression-based and a lexicon-based approach, as well as a combination of all three, in terms of article data extraction. The aim is to maximize the efficiency and accuracy of data extraction and to find out which method delivers the best results.

Our results show that the combination of all three approaches is the most powerful method to overcome the challenges of handling customer. These findings can help companies improve processes, save resources and increase customer satisfaction through faster and more accurate responses.

The paper is structured as follows. Firstly, the state of the art provides an overview of the methods used later. This is followed by an analysis of typical customer enquiries for grouping and classification. Section 4 then presents the various concepts, which are realized in Sect. 5. Section 6 then evaluates, and Sect. 7 discusses the results. The paper concludes with a summary and outlook on future research.

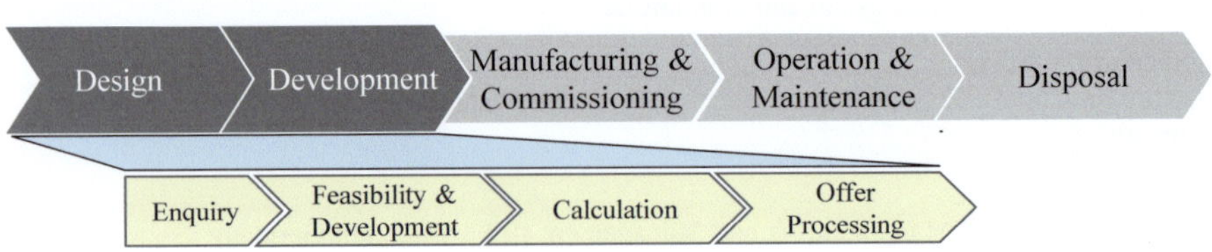


**Fig. 1** Life cycle according to Stegmaier et al. [3] and the process chain section under consideration

## 2 State of the Art

The approaches for the targeted extraction of information from texts are:

The *lexicon-based approach* uses predefined dictionaries or lexicons to recognize and categorize specific patterns and terms in texts [4]. These techniques are particularly useful for extracting structured information from unstructured texts. This approach is easier to understand and implement than machine learning [5]. For example, Sadia et al. [5] and Taj et al. [6] use this approach to find specific keywords in the text and perform sentiment analysis based on them.

The *regular expression* approach is well-established for pattern recognition and text processing, particularly in the context of information extraction [7]. The utilization of regular expressions facilitates the identification and extraction of specific strings and patterns within textual data, thereby enabling the processing of said data in a precise and efficient manner. Furthermore, they are highly customizable and can be applied to a varicty of text structures [8].

In recent years, *Large Language Models (LLMs)* have made significant advances in natural language processing [9]. These models are based on deep neural networks and large datasets and demonstrate superior capabilities for understanding and generating human language compared to previous techniques. They employ an architectural approach known as "transformer" [10]. This architecture allows contextual relationships in texts to be modelled efficiently, leading to better performance in a variety of language processing tasks, including text generation, machine translation, and question answering [11]. The use of LLMs in information extraction offers significant advantages over traditional methods. Not only can they recognize predefined patterns, but also understand contextual nuances, which increases their flexibility and accuracy. These models are particularly useful in scenarios of complex and variable text structure [11].

## 3 Analysis of Enquiry Communication of Representative Component and System Manufacturer

In conducting our analysis, we selected a representative component and system manufacturer from the vacuum handling sector. J. Schmalz GmbH provides a range of products, including individual components such as suction cups and complete gripping systems. The company's product range encompasses a total of 10,500 active sales items [12].

To classify customer enquiry types, a total of 377 existing customer enquiries were randomly selected and analyzed from the Enterprise Resource Planning (ERP) system. The emails under consideration were in the German language and the reference period for the enquiries was February to April 2024. Figure 2 illustrates one of the emails that was subjected to analysis. For reasons of confidentiality, the names

**Fig. 2** Representative email

From: Doe, John <John.Doe@doeeveprovider.de>
To: Mustermann Max - J. Schmalz GmbH
Date: Mon, 20.03.2024, 13:45
Subject: Inquiry for Vacuum Switch Schmalz

Dear Mr. Mustermann,

We kindly request a non-binding quotation for the following item:

1 piece 10.06.02.00580 VSi V D M12-4

Our customer number with you is 459841564.
Thank you very much in advance for your efforts.

Best regards,

On behalf of
John Doe
Automation Technology Consultant, Munich Branch

**Doe Eve Provider GmbH**
Hauptstraße 56 | 80336 München
Tel. +49 89 7654-321 | Fax +49 89 7654-322
John.Doe@doeeveprovider.de | www.doeeveprovider.de

Registered Office: Munich
Local Court Munich HRA 123456
Managing Director: John Doe

and other identifying information have been altered. The mail has also been translated for better understanding. In response to these enquiries, the company usually provides an offer including delivery times, availability and prices. Of the customer enquiries analyzed, a total of 92 enquiries were discarded because the content and intent of these enquiries were not clearly identifiable during the manual review, or because they were explicitly service enquiries, or complaints. The remaining 284 enquiries were classified according to the type of enquiry and assigned to one of the following categories:

- *Email*: The customer writes an email that is sent either directly to a sales representative or to the manufacturer's central communications department.
- *Website*: The customer has made a product enquiry via the manufacturer's website. For example, by using the web shop or a configurator.
- *Phone*: The customer contacts the sales representative by phone.
- *Customer visit*: The enquiry is made by a sales representative following a customer visit.
- *Trade fair*: The customer enquiry is made during a trade fair.
- *Unclear*: The origin of this enquiry cannot be clearly determined. The enquiry cannot be assigned to any of the other categories with certainty.

In accordance with the previously established categories, the analyzed customer enquiries were subsequently examined and manually assigned to one of the categories. The outcome of this analysis is illustrated in Fig. 3.

A review of the nature of the enquiries indicates that the majority of these are submitted to the company via email. This is followed by a considerable margin by enquiries received personally and by phone. In the second phase of the study, the way

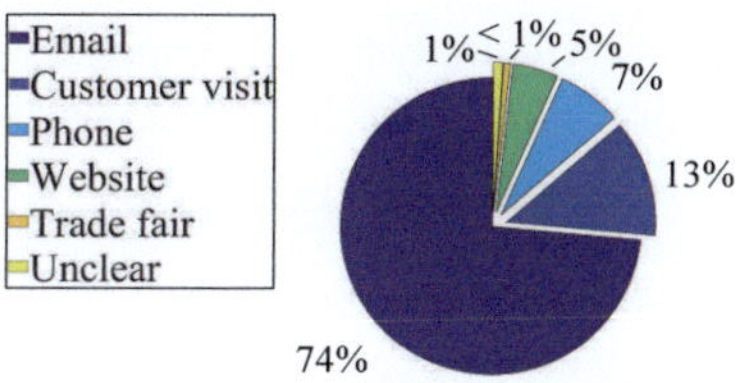


**Fig. 3** Distribution of customer enquiries by enquiry type

the content of the queries was presented was subjected to analysis and classification. The following categories were selected for the classification:

- *Enquiry with part number and/or designation*: The enquiry comprises descriptions of the articles in question, or even the specific part number of the articles that have been requested.
- *Enquiry with form*: A form is appended to the enquiry, which itemizes the requested articles with accompanying part designations and/or part numbers.
- *Enquiry with image description*: An enquiry may also be accompanied by an image description. The enquiry includes an image of the item or assembly that is the subject of the request.
- *Enquiry with questionnaire*: The enquiry is typically conducted via a standardized questionnaire, which is usually completed manually. Such a form is completed by the field service personnel during a customer visit and then forwarded to the back office for processing.
- *Enquiry for change offer*: The enquiry comprises an existing offer. It is sufficient to make only minor adaptations to the details.
- *Enquiry with application description*: The enquiry contains a description of the customer application and requests a solution to this.
- *Unclear*: It is not possible to categorize the content of the enquiry with certainty, as it does not clearly align with any of the categories.

The result of the evaluation is shown in Fig. 4. As the analysis demonstrates, most customer enquiries comprise requests that either explicitly contain a part designation or a part number, or list these in a structured format in the appendix.

Conclusion from the analysis of customer enquiries: The analysis of customer enquiries revealed that the majority of these are sent to the company by email, with the content usually already containing a specific part designation or a part number. It can be reasonably deduced that the deployment of digital tools for the automated

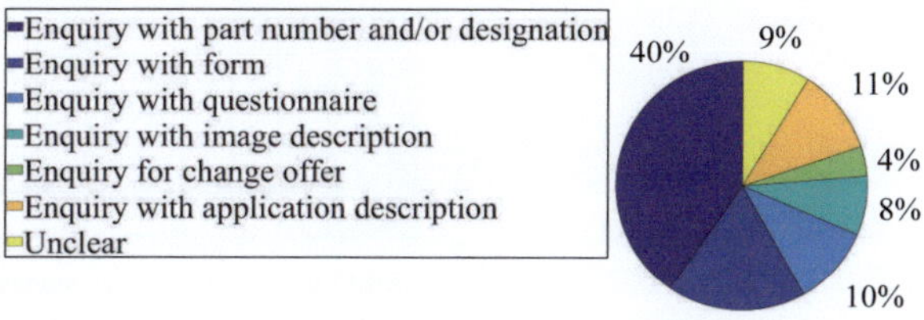


**Fig. 4** Distribution of customer enquiries by content

or assisted analysis and response to such enquiries may result in a notable reduction in the demand for human resources in sales. The subsequent sections of this paper will focus on this specific type of enquiry.

## 4 Conception

This section presents four concepts for the extraction of part designation, part number, and important context information. An example of a part number and designation is giving by:

- Part number: 10.02.02.04125
- Part designation: SCPSi 15 G02 NC M12-5.

An additional area of contextual analysis is the item quantity. In the preceding phase of the process, this information is extracted from the email. The four concepts are discussed in greater detail below.

*Concept 1* is a lexicon-based approach. The approach is based on the identification of key terms and character strings in accordance with the prevailing state of the art.

*Concept 2* is based on the use of regular expressions. The relevant information is to be extracted using expressions matched to the part designation and part numbers. To generate the expression, it is necessary to load all articles and their corresponding part numbers from the company's ERP system. To construct a comprehensive regular expression, each part designation is decomposed into individual characters and categorized by position. Unique characters in each position are then compiled into string segments. These segments are ultimately combined to form a single, regular expression capable of matching various part designations. Figure 5 illustrates the process as applied to three illustrative examples of part designations.

In contrast to the part designation, the part numbers exhibit a clearly delineated structure, which can be expressed in a more straightforward manner. This can be achieved by mapping them using a representative example of a part number and the corresponding rule chain, as illustrated in Fig. 6.

*Concept 3* is an LLM-based approach that enables rapid engineering without complex domain-specific training of the neural network. The LLM agent receives

**Figure 5** Derivation of an expression for the part designations

| Part designation | Char Position | | | | | | | | | | | | | | | |
|---|---|---|---|---|---|---|---|---|---|---|---|---|---|---|---|---|
| | 1 | 2 | 3 | 4 | 5 | 6 | 7 | 8 | 9 | 10 | 11 | 12 | 13 | 14 | … | n |
| 1 | F | S | G | C | | 4 | 0 | | N | B | R | - | 5 | 5 | … | … |
| 2 | S | A | B | | 5 | 0 | | N | B | R | - | 6 | 0 | | … | … |
| 3 | S | C | P | S | c | | 0 | 7 | | G | 0 | 2 | | N | … | … |
| ⋮ | ⋮ | ⋮ | ⋮ | ⋮ | ⋮ | ⋮ | ⋮ | ⋮ | ⋮ | ⋮ | ⋮ | ⋮ | ⋮ | ⋮ | ⋱ | ⋱ |
| r | ⋮ | ⋮ | ⋮ | ⋮ | ⋮ | ⋮ | ⋮ | ⋮ | ⋮ | ⋮ | ⋮ | ⋮ | ⋮ | ⋮ | ⋱ | ⋱ |

segment_1 = "[FS]" … segmet_m = "[…]"

part_dgn_string = segment_1 + … + segment_m

**Fig. 6** Illustration of the regular expression for part numbers

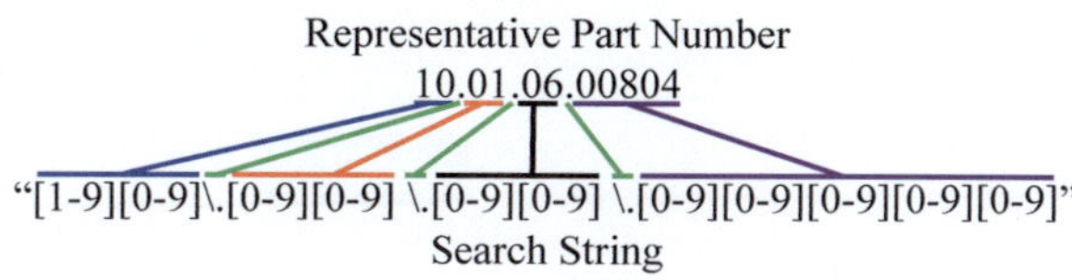


essential contextual data through prompt augmentation, following the single-turn instruction methodology [11]. The prompt is enhanced with the following information:

- **General**: “*Agent is an expert in extracting part designations, part numbers, and quantities from a text. Please sort the part designations and part numbers into the correct fields.*”
- **Part designation**: “*A part designation consists of a combination of numbers and letters. Individual sections of the part designation can be separated from each other by spaces. Example: GPQ2 122 × 87x48 SXPi/SXMPi*”
- **Part number**: “*A part number is structured as follows: XX.XX.XX.XXXXX, where X is always a number. Example: 10.02.02.02091*”
- **Quantity**: “*The number of pieces can be represented by XXx, X x, X times or X pieces, X pcs, quantity X, Menge X, X Stück, X Stk., where X is a number. However, it is also possible to specify the number in full or in another form.*”

The LLM then performs a search of the text, utilizing the enriched prompt, with the objective of extracting the requisite information, which is then outputted.

*Concept 4* combines the preceding three concepts, capitalizing on their respective advantages. It represents a hybrid approach as illustrated in Fig. 7, which depicts a comprehensive concept for the fully automated or assisted evaluation of emails with part numbers and designations. The process steps are indicated in grey and dark blue. In total, the concept comprises 14 process steps and 5 decisions. For simplicity, only the process steps marked in dark blue will be implemented in the following realization.

In the initial phase of the process, the email requests are prepared and the attachment they contain is extracted. If the email in question contains photos, these are immediately excluded from the subsequent processing phase. This is due to the necessity of utilizing a distinct and specialized toolset to extract the information from the photos in question, to facilitate the processing of the requests.

For text-based emails, the part numbers are identified using the regular expression outlined in *Concept 2*. The lexicon then verifies the validity of the identified part number and its status as an official sales item. If this is not the case, the email is marked accordingly. Furthermore, the corresponding part designation from the dictionary is linked to the part number identified.

The lexicon is then used to identify additional part designations within the text. Using the regular expression, part numbers and corresponding part designations are identified for the LLM prompt. This prompt also includes part designations identified in the previous step using the lexicon, as well as contextual information about the

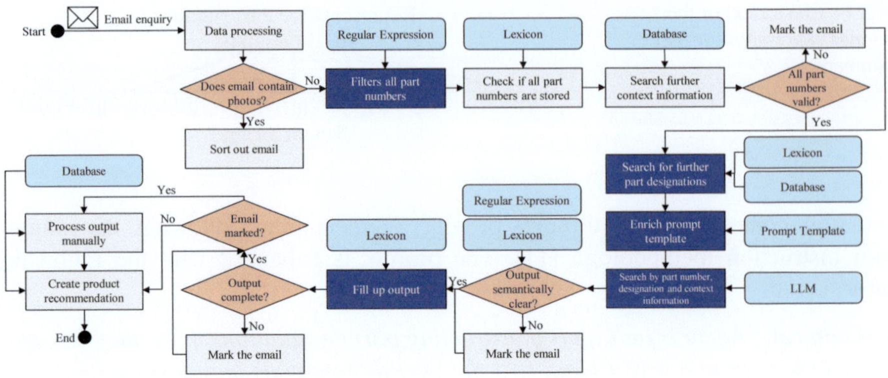


**Fig. 7** Overall process for analyzing emails using a hybrid approach

customer, such as industry information and product range content, compiled from a company-wide database.

The hybrid approach extends the LLM prompt from *Concept 3* with an additional statement:

"*The following part numbers {numbers} and part designations {designations} may appear in the text. But search for more.*"

The designation and numbers identified are employed in the construction of the placeholders for numbers and designations. The extended prompt then employs the LLM to conduct a targeted search for part numbers, part designations, and the associated contextual information (in this instance, solely the item quantity) within the textual content of the email. The output is then subjected to a semantic correctness check. For this purpose, the part numbers and part designations identified via the LLM are compared with the matches in the lexicon. If there are any discrepancies, the matches found contain a notation error or are invalid articles. Ambiguous part designations and part numbers are flagged. If the output contains flagged data or if important context information (e.g., the number of items) is missing, the email is also flagged. Subsequently, the system verifies whether the email in question was flagged at an earlier stage of the process. If this is not the case, a product recommendation can be generated automatically based on the context information and the part designations and numbers identified. Conversely, if the email was marked in a previous step, it is sorted out and the product recommendation must be prepared manually by a trained employee.

## 5 Realization

The realization of the concepts presented is briefly described in more detail below.

For *Concept 1*, part designations and part numbers of current sales articles are extracted from the ERP system and stored in a structured Excel document. The file content is then imported into a Pandas [13] data frame, creating a lexicon of all relevant entities. Each string in the email is compared against this lexicon. Matches between the email content and the lexicon entries are identified and saved in a list.

The regular expression approach of *Concept 2* is realized using MATLAB, where the "regexp" function [14] is already fully implemented. The procedure for creating the expression is also implemented.

For *Concept 3* the LangChain [15] framework is employed to actualize the LLM-based approach. In addition to facilitating the validation and serialization of the data schema, the Pydantic library is used. Pydantic provides robust validation and clear structuring of input data, enhancing the reliability and efficiency of data-intensive applications [16]. Accordingly, a data schema is established for the part number, part designation, and item quantity. This is achieved by delineating the field content and defining the data type. The LLM has the capacity to perform a total of three conversions of the output into the defined schema. If this is not feasible within the allotted three attempts, the analysis of the email is terminated. Meta's LLaMA-3:8B [17] is selected as the LLM due to its minimal memory requirements and widespread availability. The LLM is executed locally for data security reasons.

To implement the hybrid approach of *Concept 4*, the concept of regular expression is restricted to the part numbers and transferred to a python script. Nevertheless, no alterations are made to the lexicon realization described. The adjustments are implemented in the LLM. Furthermore, the approach is structured in accordance with the procedure described in Sect. 4 and the process illustrated in Fig. 7, utilizing the dark blue boxes.

## 6 Evaluation

To evaluate the four concepts, a total of 100 real and representative customer enquiries in German language were selected in the form of emails from the ERP system. The emails contain either a direct part designation or a part number. Some also contain a quantity. The data can be presented both in continuous text and in a table. For the evaluation, the data was analyzed manually and compared with the results from the various concepts.

The following scheme was developed for the evaluation: The overall score of 100% is divided evenly between the categories appearing in the email (part designation, part number, and quantity). This means that if all three categories are present, the complete fulfilment of each category corresponds to 33.33%. If there are only two categories in total, each of these categories is rated at 50%. If only one category is present, it receives 100%. The evaluation scheme always refers to all requested items. If no result is found for a category, it is given a score of 0%.

If a category is fulfilled at least halfway, it is given half of the respective score. Otherwise, it also receives a score of 0%. An excerpt from the evaluation is shown in Fig. 8.

| Mail | | Manual evaluation | | | Concept 1: Lexicon | | | | Concept 2: Regular Expression | | | | Concept 3: LLM | | | | Concept 4: Hybrid approach | | | |
|---|---|---|---|---|---|---|---|---|---|---|---|---|---|---|---|---|---|---|---|---|
| No. | Designation | p. no. | p. dgn. | qty. | p. no. | p. dgn. | qty. | sc. % | p. no. | p. dgn. | qty. | sc. % | p. no. | p. dgn. | qty. | sc. % | p. no. | p. dgn. | qty. | sc. % |
| 1 | 116625.msg | 1 | 1 | 1 | 1 | 1 | 0 | 67 | 1 | 5 | 0 | 33 | 1 | 1 | 1 | 100 | 1 | 1 | 1 | 100 |
| 2 | 111225.msg | 0 | 2 | 2 | 0 | 2 | 0 | 50 | 0 | 8 | 0 | 0 | 0 | 1 | 2 | 75 | 0 | 2 | 2 | 100 |
| 3 | 114895.msg | 2 | 0 | 2 | 2 | 0 | 0 | 33 | 2 | 4 | 0 | 33 | 2 | 0 | 0 | 50 | 2 | 2 | 2 | 100 |
| ⋮ | ⋮ | ⋮ | ⋮ | ⋮ | ⋮ | ⋮ | ⋮ | ⋮ | ⋮ | ⋮ | ⋮ | ⋮ | ⋮ | ⋮ | ⋮ | ⋮ | ⋮ | ⋮ | ⋮ | ⋮ |
| 100 | 114895.msg | 4 | 4 | 0 | 4 | 4 | 0 | 100 | 4 | 10 | 0 | 50 | 4 | 2 | 0 | 75 | 4 | 4 | 0 | 100 |
| | | | | | | Total Score | | **45** | | Total Score | | **34** | | Total Score | | **76** | | Total Score | | **87.4** |

p. no. = part number; p. dgn. = part designation; qty. = quantity; sc. = score

**Fig. 8** Extract from the evaluation of the concepts presented

*Concept 2* achieves the lowest score. All part numbers are recognized with this concept, provided they are official numbers. However, deviating or misspelled numbers are not recognized. Furthermore, there are also many false hits for the part designations, as parts of deviating strings are also recognized. The specified quantities cannot be extracted with this approach. In accordance with these limitations, the concept of regular expressions achieved a score of 34%.

*Concept 1* achieved a slightly better result of 45%. Like *Concept 2*, however, it did not achieve any results when specifying the number of items. However, it recognizes part designations more reliably. Part designations and part numbers that do not belong to an official sales item or have a spelling or syntax error are not recognized.

*Concept 3* on the other hand, can correctly extract the specified quantity from most emails. However, several incorrect character and number combinations were also found in the part designations and part numbers. Nevertheless, the concept achieved an overall score of 76%.

*Concept 4* scores highest. By using the regular expression and the lexicon, almost all part designations and part numbers are recognized correctly. Only the number of items is occasionally not recognized. This results in a score of 87.4%.

In addition to evaluating the quality of the results, the computing time required for the individual concepts was assessed. For this purpose, the time required to analyze all emails is recorded and then arithmetically averaged to determine the time per email. To determine the computing time, this process is repeated a total of three times and the mean value is calculated. Furthermore, the individual times for determining the part designation and part number are determined separately for *Concept 1* and *2*. When analyzing the calculation time, only the times for analysis were determined. Non-productive times, such as loading the lexicon, were not considered. Table 1 shows the times for analyzing all 100 emails, as well as the average computing time per email.

The mean processing time for *Concept 2* is 0.9 ms per email, making it the fastest of the concepts under consideration. The second-fastest concept is *Concept 1*.

At 98.4 ms per email, it is markedly slower than *Concept 2*, yet still considerably faster than *Concept 3* and *Concept 4*. Given the considerable size of the LLM, it is evident that *Concepts 3* and *4* are markedly slower than the other two concepts. *Concept 4* is the slowest concept, with an average processing time of 9.97 s per email, due to the combination of all three other concepts.

**Table 1** Calculation time of the four different approaches

| No | Concept | | Total time for 100 mails | Average time per mail (s) | Time comparison with employees (%) |
|---|---|---|---|---|---|
| 0 | Employees | Total | 4519* | 45.2 | 100 |
| 1 | Lexicon | p. no | 4.7714 | 0.0477 | 0.11 |
| | | p. dgn | 5.0634 | 0.0506 | 0.11 |
| | | Total | 9.8347 | 0.0984 | 0.22 |
| 2 | Regular expression | p. no | 0.0171 | 0.0002 | < 0.01 |
| | | p. dgn | 0.0682 | 0.0007 | < 0.01 |
| | | Total | 0.0853 | 0.0009 | < 0.01 |
| 3 | LLM | Total | 820.4203 | 8.2042 | 18.15 |
| 4 | Hybrid approach | Total | 997.2876 | 9.9729 | 22.06 |

p. no. = part number; p. dgn. = part designation
*Determined for 20 mails and then extrapolated

In comparison, the evaluation of the times for four employees resulted in an average time of 45.2 s per email. The employees were presented with 20 emails in a text file, which they were also required to extract the part designation, part number, and quantity from. The time required for this was recorded.

In a further test, the utilization of a larger LLM for *Concept 4* (LLaMA-3:70B [17]) resulted in a notable enhancement in the quality of the results, enabling a score of 93.4% to be achieved. Notably, the number of items is now identified with greater accuracy. However, within the same test setup, there is a notable increase in the computing time, which is attributable to the augmented memory requirements of the LLM and requires several hours.

## 7 Discussion

By using a comprehensive lexicon with all part designations and part numbers ever created in the company, the score in the evaluation of *Concept 1* can be significantly increased again. This is because all data can then be extracted from the email, if it is semantically cleanly prepared in the email, except for contextual information (e.g., number of items) and can also be clearly assigned by specifying the corresponding part number. If an incomplete part designation is given, multiple hits are still possible. In this case, the LLM approach has the advantage of identifying part designations that contain spelling errors, thereby providing this information to the subsequent editor. This can help to save time and resources when creating product recommendations.

# 8 Conclusion and Outlook

In this paper, we analyzed the content and nature of customer enquiries sent to a representative component and system manufacturer. Based on the results of the analysis, we present four different approaches for the automated extraction of relevant information from customer enquiries in emails. We focused on the part designation, the part number, and the number of items. Furthermore, we provide a comprehensive concept for processing the emails. The following results are presented in the paper, considering the component and system manufacturers:

- Most customer enquiries are in the form of an email.
- Most emails directly contain a part designation or a part number.

In general, we have found that:

- With a regular expression approach, part numbers can be extracted quickly and reliably.
- With the lexicon-based approach, many designations and numbers can be extracted from the emails.
- Using an LLM, contextual information such as the number of items can be extracted from the email.
- A hybrid approach of lexicon, regular expression, and LLM provides the best results.
- Analysis of customer enquiries based on the concepts is significantly faster than a manual analysis.

Future investigations could focus on the automated handling of exceptions or the automatic creation of product recommendations, including all information such as discount group, availability, delivery times, contact person and delivery note, using additional information from the web can also be investigated in more detail. For example, the reasoning and acting skills of LLMs (as investigated in [18]) can be transferred to the use case. It can also be investigated how the automatically evaluated data can be used to cluster the extracted information and how machine learning can be used to analyze and predict future demand for products. This allows production to be better planned and controlled.

**Acknowledgements** This work was supported by the Landesministerium für Wissenschaft, Forschung und Kunst BadenWürttemberg (Ministry of Science, Research and the Arts of the State of Baden-Wurttemberg) within the Nachhaltigkeitsförderung (sustainability support) of the projects of the Exzellenzinitiative II.

## References

1. Xiong, M.H., Tor, S.B., Bhatnagar, R., Venkat, S.: Framework to Manage Customer Enquiries for SMEs. Singapore-MIT Alliance, SMA-NTU Office, Nanyang Technological University, Singapore (2003), https://dspace.mit.edu/bitstream/handle/1721.1/3741/IMST007.pdf?sequence=2&isAllowed=y. Last accessed 28. Apr 2025/04
2. Xiong, M.H., Tor, S.B., Bhatnagar, R., Khoo, L.P., Venkat, S.: A DSS approach to managing customer enquiries for SMEs at the customer enquiry stage. Int. J. Prod. Econ. **103**(1), 332–346 (2006). https://doi.org/10.1016/j.ijpe.2005.08.008
3. Stegmaier, V., Eberhardt, T., Schaaf, W., Jazdi, N., Weyrich, M., Verl, A.: Literature Review and Model Proposal on the Machine Life Cycle in Industrial Automation from Different Perspectives. Procedia CIRP **120**, 690–695 (2023). https://doi.org/10.1016/j.procir.2023.09.060
4. Bonta, V., Kumaresh, N., Janardhan, N.: A Comprehensive Study on Lexicon Based Approaches for Sentiment Analysis. AJCST **8**(S2), 1–6 (2019). https://doi.org/10.51983/ajcst-2019.8.S2.2037.
5. Sadia, A., Khan, F., Bashir, F.: An Overview of Lexicon-Based Approach For Sentiment Analysis. In: IEEC International Electrical Engineering Conference (2018)
6. Taj, S., Shaikh, B.B., Meghji, A.F.: Sentiment Analysis of News Articles: A Lexicon-based Approach. In: 2019 2nd International Conference on Computing, Mathematics and Engineering Technologies (iCoMET), pp. 1–5, Sukkur, Pakistan (2019).
7. Bartoli, A., de Lorenzo, A., Medvet, E., Tarlao, F.: Inference of regular expressions for text extraction from examples. IEEE Trans. Knowl. Data Eng. **28**(5), 1217–1230 (2016). https://doi.org/10.1109/TKDE.2016.2515587
8. Fitzgerald, M.: Introducing Regular Expressions: Unraveling Regular Expressions, Step-by-Step, 1st edn. O'Reilly, Beijing, Köln (2012)
9. Zhao, W.X. et al.: A Survey of Large Language Models (2023) Available: http://arxiv.org/pdf/2303.18223v13. Last accessed 28 Apr 2025
10. Shanahan, M.: Talking about Large Language Models. Commun. ACM **67**(2), 68–79 (2024). https://doi.org/10.1145/3624724
11. Naveed, H., et al.: A comprehensive overview of large language models. arXiv preprint arXiv: 2307.06435 (2023). Available at: http://arxiv.org/pdf/2307.06435v9. Last accessed 28 Dec 2025
12. Fiedler, M., Schaaf, W.: Novel engineering and product solutions towards digitalization and sustainability in vacuum handling automation. In: 14th International Fluid Power Conference, pp. 888–900. Dresden, Germany (2024).
13. NumFOCUS, pandas, https://pandas.pydata.org/. Last accessed 13 Sept 2024
14. The MathWorks, Inc.: regexp: Match regular expression (case sensitive), https://de.mathworks.com/help/matlab/ref/regexp.html. Last accessed 21 Oct 2024
15. LangChain, Inc.: LangChain, https://www.langchain.com/. Last accessed 28 Apr 2025
16. Pydantic, https://docs.pydantic.dev/latest/. Last accessed 28 Apr 2025
17. Meta Platforms, Inc.: Llama 3.1 models, https://llama.meta.com/. Last accessed 13 Sept 2024
18. Yao, S., Zhao, J., Yu, D., Du, N., Shafran, I., Narasimhan, K., Cao, Y.: ReAct: Synergizing reasoning and acting in language models. arXiv preprint arXiv:2210.03629 (2022). Available at: http://arxiv.org/pdf/2210.03629v3. Last accessed 28 Apr 2025

# Go Small and Similar: A Simple Output Decay Brings Better Performance

**Cheng Xuan, Gong Haigang, Wang Xiaomin, and Liu Ming**

**Abstract** Regularization and data augmentation methods have been widely used and become increasingly indispensable in deep learning training. Researchers who devote themselves to this have considered various possibilities. But so far, there has been little discussion about regularizing outputs of the model. This paper begins with empirical observations that better performances are significantly associated with output distributions, that have smaller average values and variances. By audaciously assuming there is causality involved, we propose a novel regularization term, called Output Decay, that enforces the model to assign smaller and similar output values on each class. Though being counter-intuitive, such a small modification result in a remarkable improvement on performance. Extensive experiments demonstrate the wide applicability, versatility, and compatibility of Output Decay.

**Keywords** Regularization · Outputs · Overfitting · Data augmentation

## 1 Introduction

Improving the performance of deep convolutional neural networks (CNNs) has long been a question of great interest in a wide range of fields. Recent improvements in network structure and computer hardware have indeed brought promising performances, but they have also entailed increased risks of overfitting. To combat with this issue, a considerable literature has grown up around the themes of data augmentation and regularization. Taken together, these techniques have focused on various details of training, ranging from the inputs (Data Augmentations [1–5]), to labels(Label Smoothing [6]), to parameters (Weight Decay [7]), to depths (Stochastic Depths [8]), to neurons (Dropout [9]), to feature maps (Batch Normalization [10],

X. Cheng · H. Gong · X. Wang
University of Electronic Science and Technology of China, Chengdu, China

M. Liu (✉)
Yangtze Delta Region Institute(Quzhou), University of Electronic Science and Technology of China, Quzhou, China
e-mail: csmliu@uestc.edu.cn

 
Y. Ma (ed.), *Proceedings of 4th International Conference on Intelligent Systems Design and Engineering Applications*, Lecture Notes in Electrical Engineering 1478,
https://doi.org/10.1007/978-981-95-1802-9_2

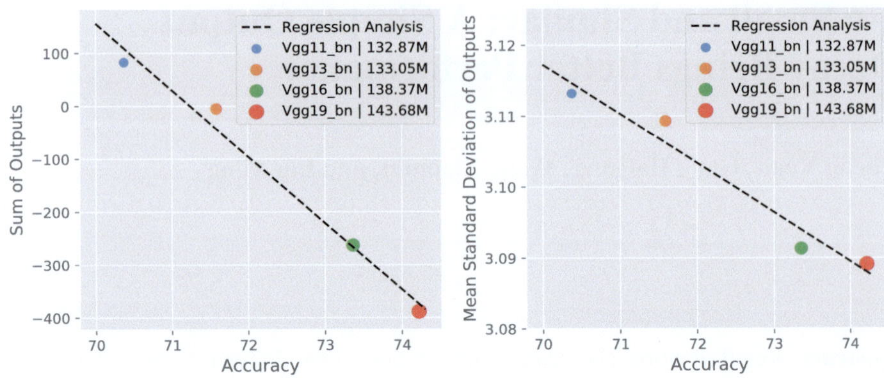


**Fig. 1** Connections between the models' performance and magnitudes(left)/variations(right) of the model's outputs. Node size represents size of parameters. The dashed lines denote the estimated results of regression analysis

DropBlock [11]) and so on. However, far little attention has been paid to regularizing the output distribution of the network.

The outputs of the network can be viewed as the projections of higher-dimensional features of images, namely, the output distribution itself implies the network's perception of images. Actually, researchers have shown an increased interest in utilizing the knowledge embedded in the output distribution. Knowledge distillation [12] believes the probabilities assigned to each class are part of the knowledge and exploit this by forcing a small network to assign the same probabilities as a large network that generalizes well. When estimating confidence of network, temperature scaling [13] that reduces all logits by a temperature factor is quite effective at calibrating a model's predictions. So it can be inferred that regularizing the output distribution should also be able to play an important role in improving the performance of the network and addressing overfitting.

To motivate our regularizer, we perform several experiments to explore whether there is a connection between the output and performance of the model. Results empirically suggest that as performances of the models increase, the mean values and variances of outputs are smaller. Consequently, we suppose it is possible to further improve the model's performance by shrinking the mean values and variances of the output distributions. In this work, we consider achieving this aim by introducing a complement loss Output Decay. This regularization term encourages the model to produce equal output value for each class. Experiments show that OD[1] achieves consistent performance gains on various benchmark datasets using multiple network architectures. Due to its inherent simplicity, this regularization term enjoys a plug-and-play nature, in other words, it can be easily assembled into any programs without worrying about it having conflict with original parts. It is hoped that this research will contribute to a deeper understanding of the effect of outputs.

[1] Throughout this paper, the term 'OD' will refer to Output Decay.

## 2 Observations

Before we delve deeper into the detailed design of our method, let us first consider a question: *is there any association between the outputs and the performance of the network? Or as commonly stated, what kind of outputs should be considered good for the model*?

To answer the question, we investigated two factors that might relate to the performance of the network: the magnitude and the standard deviation of the outputs. Specifically, we used the pretrained publicly available models provided by *Pytorch* [14]. To eliminate interference of different network structures on outputs, we chose the Vgg [15] architecture for comparison (See the Supplementary Materials for similar analytic results of ResNet and DenseNet architectures). Due to the fact that most pretrained models have suffered from overfitting on the training samples, we evaluated the models' outputs on the test samples of ImageNet [16]. We added the outputs all together to represent the magnitude of the outputs. As for the standard deviation of the outputs, we first calculated the standard deviation of each sample and then averaged them.

Figure 1 depicts how the sum and the mean standard deviation of the outputs changes as the accuracy of the models increases. Surprisingly, with increasing accuracy, the magnitude of the models' outputs shows a significant downward trend ($R^2 = 0.983$, $**p = 0.008 < .01$). The same strong and negative association exists between the mean standard deviation of the outputs and the performance of models ($R^2 = 0.956$, $*p = 0.022 < .05$).

These results suggest that, the more powerful the model is, the more regularized its outputs become. This brings us the following questions: *is it possible that the regularized outputs lead to the improvements of performance to some extent? If so, how can we utilize this in a more effective and easy-to-use way?* We will see how to address this issue next.

## 3 Method

Based on the observations and assumptions mentioned above, two properties of the outputs need to be regularized. That is to say, we should not only make outputs smaller, but also make outputs more uniformly distributed. Consequently, we propose to add an extra term to the cost function, a term called Output Decay. It is surprisingly simple. Consider $x \in R^{B\times C\times W\times H}$ and $f(x) \in R^{B\times N}$ denote a mini-batch of training images and their corresponding unnormalized outputs. $B$ and $N$ represent the mini-batch size and the number of classes in the dataset. Then, the OD term would be:

$$\mathcal{L}_{od} = \frac{1}{2} \cdot \frac{1}{B \times N} \cdot \sum_{i=1}^{B} \sum_{j=1}^{N} (f(x)_{i,j} - c)^2 \tag{1}$$

where $c$ denotes the decay level. Then the regularized cross-entropy would be:

$$\mathcal{L} = \mathcal{L}_{ce} + \beta\mathcal{L}_{od} \tag{2}$$

where $\mathcal{L}_{ce}$ represents the original cross-entropy function and $\beta$ denotes the decay coefficient.

Intuitively, the goal of our regularization term is to make each component of the outputs get closer to the decay level $c$. By presetting a small value of $c$, it not only ensures the outputs to become smaller but also guarantees the output distribution to be as much uniformly distributed as possible. This is the key of Output Decay and what we meant by *go small and similar* in the title. Having said that, large outputs can also be allowed if they considerably improve the original loss function that corresponds to the classification tasks. Put another way, our OD term can be viewed as a way of compromising between regularizing the output distribution and minimizing the original loss function. A scalar $\beta$ is used for balancing these two terms, when $\beta$ is small we prefer to minimize the original loss function, but when $\beta$ is large we push harder on regularizing outputs distribution. It may be hard to set both the decay coefficient $\beta$ and the decay level $c$ without expert knowledge on the task. We can mitigate this issue easily by treating them as hyper-parameters so that we can exhaustively evaluate the accuracy for the predefined hyper-parameter candidates with a validation dataset, which can be performed in parallel.

Output Decay is simple and incurs a negligible computational overhead. This plug-and-play regularization term can be conveniently incorporated into almost any existing deep learning projects by merely adding one line of code after evaluating the original loss function. A minimal working example with a mini-batch in PyTorch is demonstrated below to show the additional one line of code:

```
outputs = model(inputs)
ce = criterion(outputs, targets)
loss = ce + 0.5*b*pow(outputs-c, 2).mean()  # Here
   it is!
optimizer.zero_grad()
loss.backward()
optimizer.step()
```

## 4 Why Does Output Decay Help?

In order to shed some light on the effect of Output Decay, we conducted a series of systematical experiments to seek potential reasons for better generalization.

*Outputs* We first explored whether the Output Decay could make the outputs smaller as training proceeds. To this end, we counted up the absolute values of outputs and averaged them over all samples. As seen in Fig. 2, without exerting constraints on

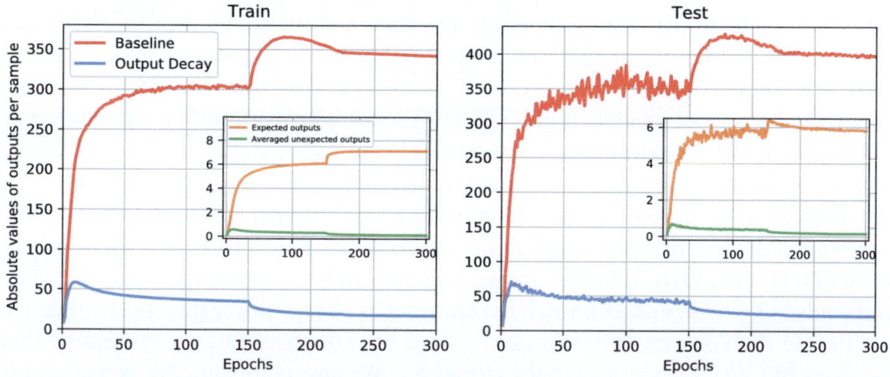


**Fig. 2** Changes of absolute values of outputs as training(left)/testing(right) proceeds. **Center Right Insets**: Comparison between absolute values of expected outputs and averaged unexpected outputs of OD. These results show while OD shrinks the magnitudes of outputs, the expected outputs can still get larger due to the original loss function

**Fig. 3** Reliability diagram for CIFAR-100 with ResNet-110 trained with and without OD. This graph demonstrates that OD helps produce well-ranked confidence estimates

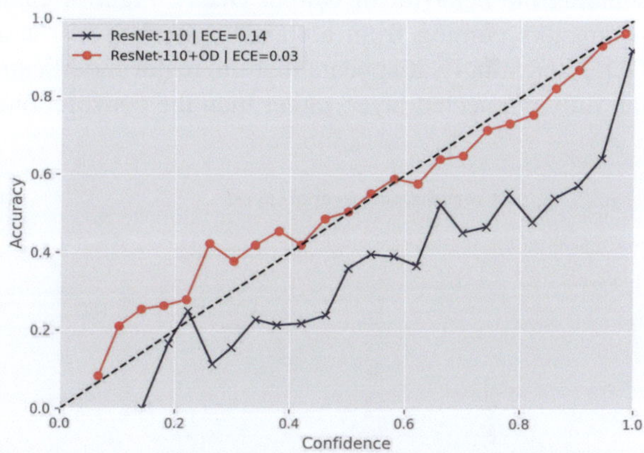


outputs, the outputs in baseline training became larger and larger during training. This trend stays the same while testing. Conversely, OD indeed shrinks outputs, as expected. We then delved deeper into the behavior into Output Decay. The insets in Fig. 2 depict the change of both expected (outputs that correspond to the ground-truth label) and averaged unexpected (sum of outputs that corresponds to other labels and divided by the number of those labels) outputs, produced by models trained with OD during training and testing. It can be easily observed that whereas the mean value of unexpected outputs takes on a decreasing change due to the regularized effect of OD, the expected outputs still manage to rise so that models can still maintain comparable representation abilities. These results are further proofs of the "compromising" behavior of OD mentioned earlier.

*Calibration* The above 'compromising' phenomenon makes us wonder if OD can improve the calibration of the model by making the confidence of its predictions more accurately represent their accuracy. Calibrated confidence estimates are very important for model practicability, since a model with bad quality of confidence estimates is unreliable and can not be trusted with on some safety-critical applications such as autonomous driving or computer-aided diagnosis. To seek the answer to the question, we investigate the calibration of the model trained with and without OD.

Figure 3 depicts the 25-bin reliability diagrams of ResNet-110 trained with and without OD on CIFAR-100. We can see that the vanilla model is quite un-calibrated and tends to be overconfident since the accuracy is always lower than the confidence values. On the contrary, we then can observe the effects of OD on calibration that the slope of OD is very closer to the slope of 1. We also computed the expected calibration error (ECE) [17]. Again, OD achieves a better result. It is quite remarkable since such a simple method can lead to such well-ranked confidence estimates.

*Parameters* Observations on changes of parameters are of vital importance in understanding the behavior of Output Decay. Figure 4 characterizes the distributions of parameters coming from a standard ResNet-110 and a ResNet-110 trained with OD. Interestingly, it appears that the major modification to parameters happens on the fully-connected layer, rather than the convolutional layers. By regularizing the

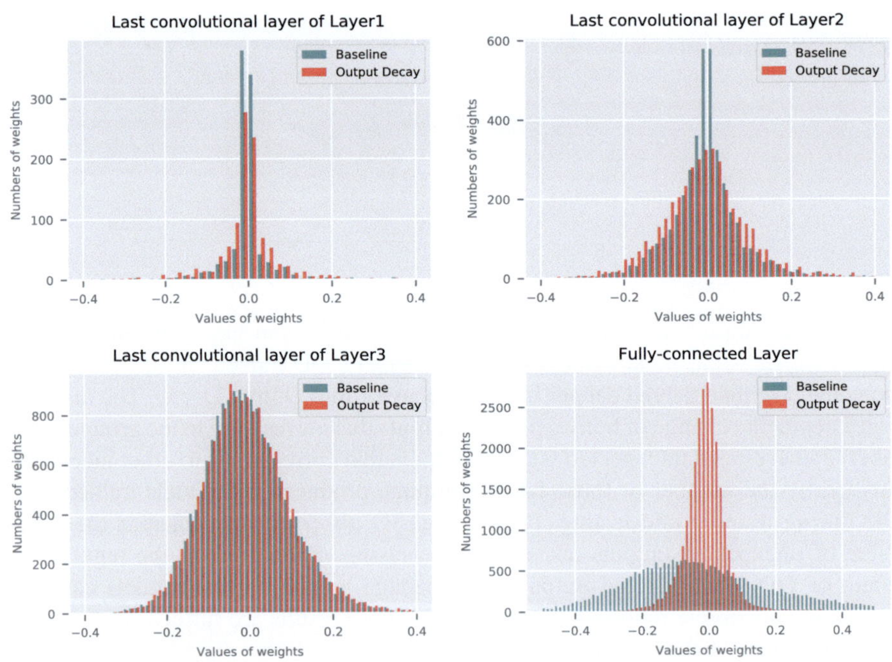


**Fig. 4** The distributions of parameters. A standard ResNet-110 ($Acc = 74.04\%$) is compared with a ResNet-110 trained with OD ($Acc = 76.09\%$) at four different layers

output distributions, OD undercuts most large weights, which suggests that OD helps provide a simpler and more powerful explanation for the data. In addition, the smallness of weights brings the model an ability to resist underlying noise information of data. A network with smaller weights prefers to learn from evidence which is seen often across the training set while a network with large weights may change its behavior frequently in response to some specific information. Detailed empirical experiments of this effect will be shown in the following contents.

*Activations of Feature Map* To investigate how feature information flows through the network while applying Output Decay, we then visualize the feature activations after the last convolutional layer of the network (See the Supplementary Materials for similar analytic results of feature maps from other layers). Specifically, we randomly selected a testing sample and extracted its corresponding feature map ($256 \times 8 \times 8$), denoted by $F^{C \times H \times W}$. For better illustration, we superimposed the statistics of 256 partial feature maps ($8 \times 8$) into a matrix $M^{H \times W}$. Since there were a large number of inactive points in the feature map, we chose to only count the number of points with values greater than 0. We can obtain that:

$$M_{i,j} = \sum_{c=1}^{C} \mathbb{1}(F[c, i, j] > 0) \tag{3}$$

where $\mathbb{1}$ denotes the indicator function and it tests a conditional expression. Then we plotted this matrix $M$ into a 3-D plane, where x-axis and y-axis denote $i$ and $j$, respectively. We used z-axis to denote the value of corresponding points, namely, each bar in Fig. 5 represents how many times these 256 partial feature maps have been activated at that position. We also provide the results obtained from the network trained with a regular scheme.

As can be seen from the graph, while both demonstrating similar activation situations, there are fewer points being activated in the feature map from the network trained with OD. On one hand, this sparse activation situation symbolizes a stronger *feature selection* ability, on the other hand, the sparse activation makes the network become more difficult to fit the data, meaning that OD also *suppresses the emergence of overfitting* to a certain extent.

*Visualizations of Feature Distributions* In order to intuitively understand the effect of Output Decay on the network, we perform a visualization scheme similar to [18]. It is based on three steps: (1) Add a new fully-connected layer between the last convolutional layer and the original fully-connected layer. The output number of this layer is 2 so that we can directly plot its features onto a plane. (2) Fine-tune the input number of the original fully-connect layer. (3) Train and project all the outputs of this new layer onto a 2-D plane. Figure 6 compares the visualization results obtained from the networks trained with and without OD.

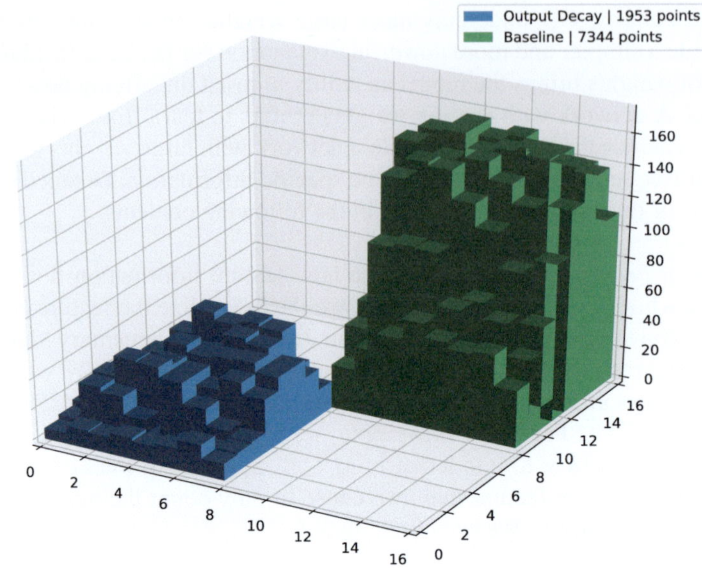


**Fig. 5** Visualizations of activation situation. The size of blue/green area is $8 \times 8$. The height ($z$) of the bar represent the number of times the point at this position ($x - y$) is activated. In a feature map with a total of 16384 points, OD activated 1953 points and baseline (without OD) activated 7344 points. The results present that the activation of network trained with OD is more sparse

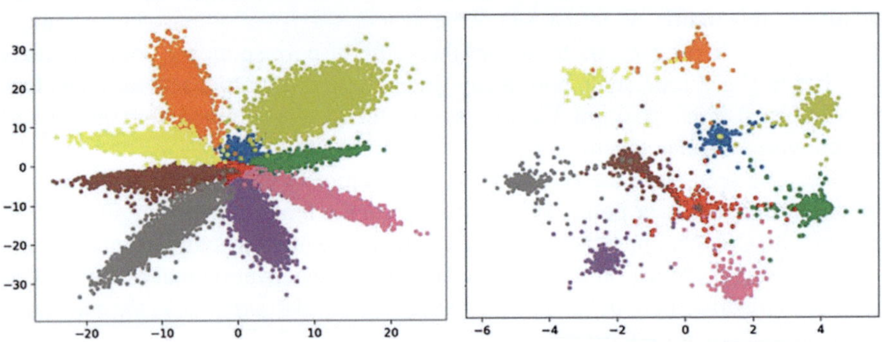

**Fig. 6** Visualizations of 2-D feature distribution on ResNet-110 on CIFAR10. Left plot: Trained without OD; Right plot: Trained with OD. Results show that OD helps obtain more discriminative features

This graph is quite revealing in several ways: (1) under the supervision of cross-entropy loss, the deeply learned features are separable, but (2) they are not discriminative enough, as evidenced by the significant intra-class variations. By contrast, (3) the projections of OD are spread into 10 defined and tight clusters, demonstrating that OD results in better discriminative feature learning in terms of smaller intra-class variations and larger inter-class distances.

# 5 Experiments

## *5.1 Output Decay Can Be Universally Applied to Various Architectures*

We first extensively evaluate Output Decay on various architectures: Vgg [15], MobileNets [19, 20], ResNet [21], preResNet [22], ResNext, WideResNet [23], DenseNet [24], and PyramidNet [25]. All the experiments were performed on CIFAR-100 [26] that contains 50000 training and 10000 testing images divided into 100 classes. For fair comparison, all the experiments strictly use the same preprocessing and data augmentation strategies such as random flipping and cropping, following the common practice. All the hyper-parameters and architectural details remain unchanged too (See Supplementary Materials for full implementation details). All experiments were conducted four times and the averaged best performances during training are reported.

Notably, the results in Table 1 show that OD consistently offers significant performance gains over the baselines, such as an impressive 1.37% top-1 error reduction on ResNet-110. We would like to note the *practical and theoretical implications* provided by Output Decay. Practically, any CNN-based architecture can acquire performance gains through negligible adjustments. Theoretically, the fact that such a small alteration results in such a significant improvement could offer much food for thought regarding the nature and extent of output regularization.

**Table 1** Top-1 error (%) with and without OD using different architectures on CIFAR-100

| Model | Vanilla | + Ours |
|---|---|---|
| Vgg19_bn | $28.45 \pm 0.11$ | **27.82** $\pm 0.05$ |
| MobileNet | $31.54 \pm 0.17$ | **30.27** $\pm 0.15$ |
| MobileNetV2 | $27.33 \pm 0.12$ | **26.81** $\pm 0.09$ |
| ResNet-56 | $26.34 \pm 0.01$ | **25.08** $\pm 0.13$ |
| ResNet-110 | $25.47 \pm 0.09$ | **24.1** $\pm 0.02$ |
| ResNet-56-PreAct | $25.82 \pm 0.06$ | **25.07** $\pm 0.01$ |
| DenseNet-100-12 | $22.78 \pm 0.04$ | **21.72** $\pm 0.03$ |
| WRN-28-10 | $18.79 \pm 0.05$ | **18.43** $\pm 0.06$ |
| ResNext-8-64 | $18.09 \pm 0.17$ | **17.51** $\pm 0.06$ |
| PyramidNet-110-270$_{\text{(BasicBlock)}}$ | $18.63 \pm 0.10$ | **17.92** $\pm 0.08$ |
| PyramidNet-200-240$_{\text{(Bottleneck)}}$ | $17.09 \pm 0.13$ | **16.61** $\pm 0.16$ |

**Table 2** Top-1 error (%) with and without OD on multiple benchmark datasets

| Datasets | Model | Vanilla | + Ours |
|---|---|---|---|
| Fashion-MNIST | ResNet-29 | 4.83 $\pm$ 0.02 | **4.64** $\pm$ 0.01 |
| SVHN | ResNet-29 | 4.3 $\pm$ 0.03 | **3.82** $\pm$ 0.02 |
| CIFAR-10 | ResNet-29 | 7.16 $\pm$ 0.06 | **6.68** $\pm$ 0.06 |
| Tiny-ImageNet | ResNet-110 | 37.62 $\pm$ 0.04 | **36.04** $\pm$ 0.11 |
| ImageNet | ResNet-50 | 23.77 $\pm$ 0.02 | **22.68** $\pm$ 0.01 |

## *5.2 Output Decay Is Equally Effective on Multiple Benchmark Datasets*

We then extended the experiments to other benchmark datasets. In particular, we used the following five benchmark datasets: Fashion-MNIST [27], SVHN [28], CIFAR-10 [26], Tiny-ImageNet, ImageNet [16]. Together with the previous CIFAR-100, we believe these six datasets should be able to cover a wide range of scenarios in the computer vision field. The implementation details of these datasets can be found in Supplementary Materials.

We show the results in Table 2. These results prove the wide applicability on various datasets. We draw attention to the fact that OD performs even better on larger and harder datasets, which is not surprising as the potential of models trained on small datasets has nearly been drained out.

## *5.3 Output Decay Is Compatible with Various Techniques*

In practice, scenarios in which more than one regularization or data augmentation techniques are used occur frequently. Therefore, compatibility with other techniques is also an important indicator of practical implication. Hence, we further apply Output Decay with some existing commonly used techniques, such as Cutout [2], Mixup [29], Autoaugment [3], FashAutoAugment [4], Label Smooth [6], and Confidence Penalty [30]. To ensure the effectiveness of the proposed method, a strong baseline is used here, which is PyramidNet-110 with widening factor $\alpha = 270$. The detailed implementation settings are described in Supplementary Materials.

It can be seen from the data in Table 3 that OD is able to make further improvements, even on some strong techniques. We want to note again that though these improvements are not as high as one might expect, the improvements due to OD come at zero architectural alteration, nearly one line of code modification, negligible memory use and computational time increase.

**Table 3** Top-1 and Top-5 error (%) of PyramidNet-110-270 (BasicBlock, the top-1 error of the vanilla model is 18.63%) with the combination of OD and some commonly used techniques on CIFAR-100

| Model | Method | Top-1 | Top-5 |
|---|---|---|---|
| PyramidNet | +Cutout | 17.52 | 4.01 |
| | +Cutout+OD | **16.94** | 4.20 |
| | +Mixup | 16.90 | 4.35 |
| | +Mixup+OD | **16.53** | 4.18 |
| | +AutoAugment | 17.20 | 3.68 |
| | +AutoAugment+OD | **16.16** | 3.72 |
| | +FastAutoAugment | 17.43 | 3.60 |
| | +FastAutoAugment+OD | **16.66** | 3.83 |
| | +LabelSmooth | 18.56 | 5.13 |
| | +LabelSmooth+OD | **18.22** | 5.05 |
| | +ConfidencePenalty | 18.41 | 4.92 |
| | +ConfidencePenalty+OD | **17.90** | 4.68 |

Results demonstrate the wide compatibility of OD

## 6 Discussion

We have introduced Output Decay that regularizes outputs to improve the performance and generalization of networks. It's simple that it can be conveniently incorporated with CNNs without any other change entailed. Experiments empirically demonstrated that OD can be applied to various architectures, datasets, and classification scenarios, including confidence estimates, label noise, and small data regime.

Despite its exploratory nature, this study offers some insight into how to operate on outputs and why brings improvements. It is hoped that this work may generate fresh insight into deep learning.

**Acknowledgements** This work was supported by the National Key Research and Development Program under Grant 2022YFB3104600, the Municipal Government of Quzhou (Grant 2023D007, Grant2023D014, Grant 2023D033,Grant 2023D034, Grant 2023D035), and Guiding project of Quzhou Science and Technology Bureau (2022K50, 2023K013 and 2023K016).

## References

1. Yun, S., Han, D., Oh, S.J., Chun, S., Choe, J., Yoo, Y.: Cutmix: regularization strategy to train strong classifiers with localizable features. In: International Conference on Computer Vision (ICCV) (2019)
2. DeVries T, Taylor GW (2017) Improved regularization of convolutional neural networks with cutout. arXiv preprintarXiv:1708.04552 (2017)
3. Cubuk ED, Zoph B, Mane D, Vasudevan V, Le QV (2018) Autoaugment: learning augmentation policies from data. arXiv preprintarXiv:1805.09501 (2018)

4. Lim, S., Kim, I., Kim, T., Kim, C., Kim, S.: Fast autoaugment. In: Advances in Neural Information Processing Systems, pp. 6665–6675 (2019)
5. Zhong, Z., Zheng, L., Kang, G., Li, S., Yang, Y.: Random erasing data augmentation. In: AAAI, pp. 13001–13008 (2020)
6. Szegedy, C., Vanhoucke, V., Ioffe, S., Shlens, J., Wojna, Z.: Rethinking the inception architecture for computer vision. I: Proceedings of the IEEE Conference on Computer Vision and Pattern Recognition, pp. 2818–2826 (2016)
7. Krogh, A., Hertz, J.A.: A simple weight decay can improve generalization. In: Advances in Neural Information Processing Systems 4, vol. 4, pp. 950–957 (1991)
8. Huang, G., Sun, Y., Liu, Z., Sedra, D., Weinberger, K.Q.: Deep networks with stochastic depth. In: European Conference on Computer Vision, pp. 646–661. Springer (2016)
9. Srivastava, N., Hinton, G., Krizhevsky, A., Sutskever, I., Salakhutdinov, R.: Dropout: a simple way to prevent neural networks from overfitting. J. Mach. Learn. Res. **15**(1), 1929–1958 (2014)
10. Ioffe, S., Szegedy, C.: Batch normalization: accelerating deep network training by reducing internal covariate shift. In: Proceedings of The 32nd International Conference on Machine Learning, pp. 448–456 (2015)
11. Ghiasi, G., Lin, T.Y., Le, Q.V.: Dropblock: a regularization method for convolutional networks. In: Advances in Neural Information Processing Systems, pp. 10727–10737 (2018)
12. Hinton, G.E., Vinyals, O., Dean, J.: Distilling the knowledge in a neural network. arXiv preprintarXiv:1503.02531 (2015)
13. Guo, C., Pleiss, G., Sun, Y., Weinberger, K.Q.: On calibration of modern neural networks. In: Proceedings of the 34th International Conference on Machine Learning, vol. 70, pp. 1321–1330 (2017)
14. Paszke, A., Gross, S., Massa, F., Lerer, A., Bradbury, J., Chanan, G., Killeen, T., Lin, Z., Gimelshein, N., Antiga, L., et al.: Pytorch: an imperative style, high-performance deep learning library. In: Advances in Neural Information Processing Systems, pp. 8026–8037 (2019)
15. Simonyan, K., Zisserman, A.: Very deep convolutional networks for large-scale image recognition. arXiv preprintarXiv:1409.1556 (2014)
16. Russakovsky, O., Deng, J., Hao, S., Krause, J., Satheesh, S., Ma, S., Huang, Z., Karpathy, A., Khosla, A., Bernstein, M., et al.: Imagenet large scale visual recognition challenge. Int. J. Comput. Vision **115**(3), 211–252 (2015)
17. Naeini, M.P., Cooper, G.F., Hauskrecht, M.: Obtaining well calibrated probabilities using Bayesian binning. In: AAAI'15 Proceedings of the Twenty-Ninth AAAI Conference on Artificial Intelligence, vol. 2015, pp. 2901–2907 (2015)
18. Wen, Y., Zhang, K., Li, Z., Qiao, Y.: A discriminative feature learning approach for deep face recognition. In: European Conference on Computer Vision, pp. 499–515. Springer (2016)
19. Howard, A.G., Zhu, M., Chen, B., Kalenichenko, D., Wang, W., Weyand, T., Andreetto, M., Adam, H.: Mobilenets: efficient convolutional neural networks for mobile vision applications. arXiv preprintarXiv:1704.04861 (2017)
20. Sandler, M., Howard, A., Zhu, A., Zhmoginov, A., Chen, L.C.: Mobilenetv2: inverted residuals and linear bottlenecks. In: Proceedings of the IEEE Conference on Computer Vision and Pattern Recognition, pp. 4510–4520 (2018)
21. He, K., Zhang, X., Ren, S., Sun, J.: Deep residual learning for image recognition. In: Proceedings of the IEEE Conference on Computer Vision and Pattern Recognition, pp. 770–778 (2016)
22. He, K., Zhang, X., Ren, S., Sun, J.: Identity mappings in deep residual networks. In: European Conference on Computer Vision, pp. 630–645. Springer (2016)
23. Zagoruyko, S., Komodakis, N.: Wide residual networks. arXiv preprintarXiv:1605.07146 (2016)
24. Huang, G., Liu, Z., Van Der Maaten, L., Weinberger, K.Q.: Densely connected convolutional networks. In: Proceedings of the IEEE Conference on Computer Vision and Pattern Recognition, pp. 4700–4708 (2017)
25. Han, D., Kim, J., Kim, J.: Deep pyramidal residual networks. In: Proceedings of the IEEE Conference on Computer Vision and Pattern Recognition, pp. 5927–5935 (2017)

26. Krizhevsky, A., Hinton, G., et al.: Learning multiple layers of features from tiny images (2009)
27. Xiao, H., Rasul, K., Vollgraf, R.: Fashion-mnist: a novel image dataset for benchmarking machine learning algorithms (2017)
28. Netzer, Y., Wang, T., Coates, A., Bissacco, A., Wu, B., Ng, A.Y.: Reading digits in natural images with unsupervised feature learning (2011)
29. Dauphin David Lopez-Paz, Y.N., Zhang, H., Cisse, M.: mixup: beyond empirical risk minimization. In: International Conference on Learning Representations (2018)
30. Pereyra, G., Tucker, G., Chorowski, J., Kaiser, Ł., Hinton, G.: Regularizing neural networks by penalizing confident output distributions. arXiv preprintarXiv:1701.06548 (2017)

# Security Framework Development for Adaptive Phishing Detection Using Hybrid Deep Learning Techniques

**Mosiur Rahaman, Nicko Cajes, Brij B. Gupta, and Kwok Tai Chui**

**Abstract** Phishing is an online attack that takes advantage of victims' technological ignorance, occasionally including a Uniform Resource Locator (URL). Phishing websites are the primary target for attackers seeking to invade individuals' personal addresses. Many phishing attack detection solutions have been created; however, many have either been ineffective or lack the compact attributes necessary for practical use. To address this issue, innovative techniques are employed, such as Hybrid Deep Learning (DL) Techniques, which combine CNN and LSTM models to accurately comprehend the intrinsic properties of phishing URLs and recognize them. With an overall accuracy of 99.99%, we experimentally proved on standard datasets that the combined identity detection approach greatly enhances phishing detection capacity. It has been demonstrated that the suggested hybrid deep learning technique is useful addition that will provide the phishing prevention ecosystem with better defense measures over rapidly evolving methods used by phishing.

**Keywords** Phishing detection · Deep learning · Security · URL · Spear phishing

M. Rahaman · N. Cajes · B. B. Gupta (✉)
International Center for AI and Cyber Security Research and Innovations, Asia University, Taichung, Taiwan
e-mail: gupta.brij@gmail.com

N. Cajes
Northern Bukidnon State College, Manolo Fortich, Philippines

B. B. Gupta
Computer Science and Information Engineering, Asia University, Taichung, Taiwan

VIZJA University, Warsaw, Poland

K. T. Chui
Hong Kong Metropolitan University, Hong Kong SAR, China
e-mail: jktchui@hkmu.edu.hk



Y. Ma (ed.), *Proceedings of 4th International Conference on Intelligent Systems Design and Engineering Applications*, Lecture Notes in Electrical Engineering 1478,
https://doi.org/10.1007/978-981-95-1802-9_3

## 1 Introduction

Phishing attacks have emerged as a notorious type of cyber threat becoming well-known with its sophisticated attack techniques due to technological advancements over the last century since its identification in 1996 [1]. Though the emergence of technology is promising for users, it has also incited cyber attackers to initiate more complex cyber-attacks. So, over the years, various phishing attacks have been identified such as spear phishing, smishing, vishing, and whaling, each utilizing different strategies and different method of attack [2].

A typical phishing attack starts with an attacker sending a message with malicious links that contains content that sounds too good to be true or contain alerts that drive the victim into a panic (positive or negative) [3], such that he clicks the link and is redirected to a fake website that perfectly imitates legit ones [4]. Often, these malicious websites require some personal information of the victim (usernames, bank credentials), which are collected by the attacker for potential identity theft [5] or to access bank account information causing huge financial loss. This emerging problem has thus become a threat to not only individuals but also to organizations, highlighting the necessity of developing a robust security system. Accordingly, researchers in the cyber security field have developed a variety of innovative solutions that can mitigate and solve the problem [6].

One recent innovative approach has been the involvement of Artificial Intelligence (AI) as a potential factor that can resolve the problem using AI/ML (machine learning) /DL (deep learning) techniques [7]. Considering the unique ability of AI techniques to identify different patterns (normal and abnormal) with the help of training from previous data [8], it has emerged as a top option next to traditional security measure approaches.

Accordingly, the present paper presents a novel approach that uses hybrid DL techniques, combining Convolutional Neural Network (CNN) and Long Short-Term Memory (LSTM) models to accurately comprehend the intrinsic properties of phishing URLs to detect phishing attacks. The rest of the paper is structured as follows: Sect. 2, presents the literature review of recent studies in phishing attack detection, Sect. 3 briefly discusses our proposed methodology, Sect. 4 discusses and analyzes the experimental result along with a comparison of our proposed result with related studies, and finally Sect. 5 presents an overall summary of the study along with future implications.

## 2 Related Work

Prior research has proposed different ways of detecting phishing attacks by analyzing URLs, emails, and attacks on IoT ecosystem. Among this, the emergence of AI has immensely helped in detecting website phishing attacks. The most common classifiers in detection have been DL [9] and ML classifies such as Random Forest and Naïve Bayes algorithm [10]. However, the persistence of XSS encryption attacks [11] and malicious URLs necessitates the implementation of innovative solution.

In this light, the use of ensemble deep learning framework that improve URL extraction and phishing detection was conducted by [11], which utilize dynamic feature selection process, optimal feature selection, and combination of different AI techniques. Study done by [12] developed a phishing mail detection system using ensemble classification model and achieved great result which is higher than other models by combining the strengths of SVM, XGB, and LR in a stacking approach. This highlights the superiority of ensemble learning methods over traditional ones and demonstrates how effective they are at classifying phishing emails. In a comparable way, [13] suggested HELPHED, another hybrid method for detecting phishing email. With an F1-score of 99.42%, the soft voting ensemble learning outperforms well-known ML/DL algorithms and previous works. HELPHED's stacking and soft voting ensemble learning method, each use a distinct ML algorithm to separately handle the hybrid features. Their experiment demonstrates that using hybrid features in conjunction with ensemble learning improves detection performance when compared to using the content-based features or text-based features.

In general, AI techniques is realistically helpful innovation in detecting phishing attacks in different domain. Advancements of methods like ensemble techniques, explainable AI [11] becomes one of its interesting parts specially as a useful feature selection method for phishing detection based on DL [14] and ML, can also be integrated. A study of [15] took a direct approach, implementing explainable AI in cyber-risk management which is a hybrid architecture for phishing attack defense and recovery, through its excellent result of true positive 92.35%, true negative of 96.4%, precision of 92.35%, and recall of 92.3%, it highlights the effectiveness of explainable AI in tackling this problem. In addition DL based intelligent cyber-security phishing detection solution was developed by [16] which effectively enhance outcomes and record phishing and non-phishing emails with consecutive accuracy rates of 88%, 100%, and 97%. An innovative hybrid DL model for cyber-security threat identification that incorporates an optimization algorithm was developed by [17], by combining state-of-the-art deep learning and optimization techniques. Their model is capable of detecting data and applications infected with malware outperforming traditional approaches in terms of categorization by achieving an excellent accuracy result. A somewhat different approach was developed by [18], which is also a hybrid phishing detection method but uses a joint visual and text identity. This technique makes advantage of unique visual characteristics that resemble human eyesight to improve the accuracy in detecting the logo. Their method has shown impressive results on many benchmark datasets, with an overall accuracy of 98.6%, proving its practical and significant contribution to the anti-phishing community.

Another hybrid machine learning approach based on static and dynamic analysis was carried out by [19]. The suggested technique can process any Android application and applies state-of-the-art ML algorithms like SVM, KNN, and various ensemble methods to assess the execution of all classifiers, which achieves a high accuracy of 100%, outperforming cutting-edge techniques. To combat the growing threat of phishing attacks and support digital forensics, [20] developed an additional improved phishing detection model for cybercrime forensics employing a cutting-edge hybrid DL framework achieving an accuracy rate of 98%, representing a significant advancement in the ongoing fight against phishing. One phishing attack specifically the spear phishing, was the focus of and developed machine learning-based spear phishing attack defenses using the ML algorithm to analyze various features like email contents, sender behavior, and attachment to detect potential threats. Among their tested models, with a 99.99% accuracy rate, the CNN for hierarchical models was the most accurate having a result of almost perfect and the sequential Bi-LSTM comes next with 99.94%. In comparison, the conventional Multi-Layer Perceptron (MLP) demonstrated a 98.46% accuracy rate, highlighting the superiority of sophisticated sequential hierarchical models over conventional methods in spear phishing assault detection.

## 3 Proposed Approach

In this section, we discuss about our proposed approach, highlighting data collection and reprocessing. The proposed model combining CNN and LSTM involves six stages that starts with the input layer and ends with the output layer. Also discussed are the different strategies, training process, validation and testing method.

### *3.1 Data Collection and Preprocessing*

The data collected comes from a publicly available tool for research and development in phishing detection systems, the Phi-USIIL Phishing URL Dataset Carefully chosen to give a balanced representation for efficient model training and evaluation, this dataset contains 134,850 legitimate URLs and 100,945 phishing URLs [https://www.kaggle.com/datasets/sharmageetika/phishing-url-dataset]. The collection mostly consists of current and active URLs, so it provides sensitivity to current phishing and appropriate website traits.

Advanced feature engineering and thorough study are made possible by extracting features from the source code of web pages as well as the URL structure. Advance feature engineering was conducted throughout the process, including the measurement of character continuation patterns in URLs. In addition, we assess the correspondence between the URL and the webpage name and estimates the chance of character occurrence in a URL based on observed frequencies and evaluates the reliability possibility of the top-level domain. The Top 10 most important features of our selected dataset are shown in Fig. 1.

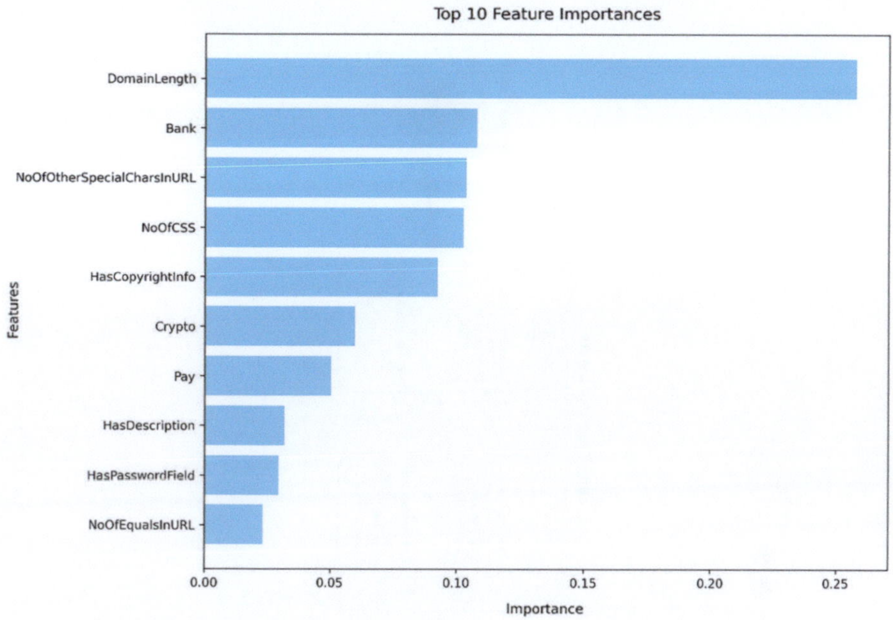


**Fig. 1** Top 10 important features of the selected dataset

## 3.2 Proposed Framework and Model Development

The development of the hybrid deep learning model which combines the CNN and LSTM networks have come in a lot of process starting from the input layer, CNN component, LSTM Component, combination of CNN and LSTM outputs, and the Output Layer. The adaptive mechanisms obtained through the careful selection and combination of deep learning models in our proposed framework further enhanced its ability to detect evolving phishing attacks and dynamically learn from the ever-evolving phishing tactics. With its effectiveness in fine-tuning the parameters of our model using the training data in real time, our model has become essential and achieved an excellent result. The model development framework of our proposed model presenting the major components of its classification performance is shown in Fig. 2.

The first processing stage starts with the input layer which is expecting a 3D input shape and is consisting of the number of timesteps and features at each time step. The development of CNN component was then followed to capture spatial patterns which contains six layers namely Conv1D, MaxPooling1D, Batch Normalization, second Conv1D, MaxPooling1D, and Flatten Layer. For the first ConV1D layer, 128 filters with the kernel size of 5 was applied together with the utilization of ReLU activation function for non-linearity, it was then followed by a MaxPooling1D layer with a pool size of 12 to reduce the spatial awareness and a Batch Normalization layer for normalizing the activations. The second Conv1D layer then applies 64 filters

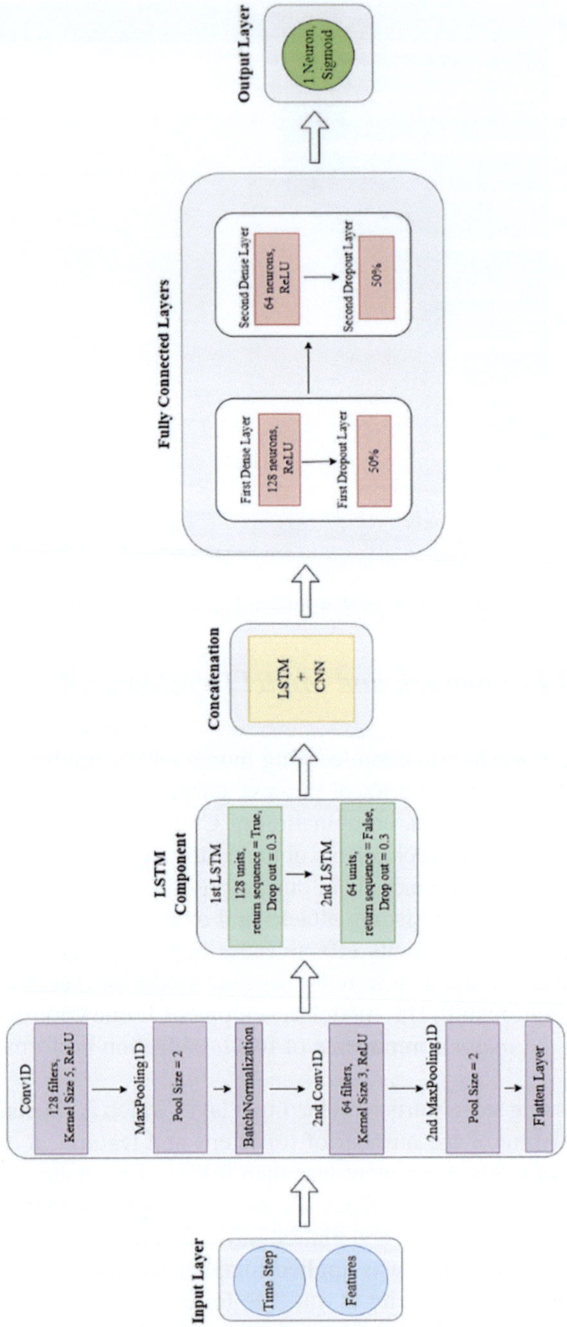


**Fig. 2** CNN and LSTM model development framework

with a kernel size of 3 together with the utilization of the same activation function, it was then followed by another MaxPooling1D Layer with pool size of 2 for further dimensionality reduction. Lastly, the flatten layer then converts the CNN output into a 1D vector to be fed to dense layer.

The next stage prioritized on the capturing of temporal dependencies in the sequential data utilizing the component of LSTM which has two LSTM layers. The first LSTM layer employs 128 units and returns the sequences to allow its output to be passed in the second LSTM layer which incorporates a dropout of 30% for regularization. The second LSTM layer contains 64 units with a return sequence that is set to false, which will produce the final output for the LSTM branch while also applying the dropout of 30% for regularization furtherly.

Following that, the output coming from CNN and LSTM branches are being concatenated into one vector. This will effectively merge the spatial features extracted with temporal features by the CNN with the temporal features captured by the LSTM to represent it in a unified form. After that, the fully connected layers then refine the extracted features, which includes two dense and dropout layers. For the first dense layer, 128 neurons using the activation function of ReLU was utilized followed by a dropout layer with 50% rate to reduce overfitting. For the second dense layer, 64 neurons with the same activation function of ReLU was utilized accompanied with another dropout layer of 50% to regularize the model furtherly.

The final processing layer is the output layer. This layer contains a single neuron with sigmoid activation function, which aids the model to give an output of probability between 1 and 0. In Overall, the model utilized the Adam optimizer in compilation together with a learning rate of 0.0001, binary cross-entropy is then applied for the loss function as it is appropriate for binary classification problems and accuracy metrics is used to effectively monitor the performance of the model.

## 3.3 Selection of Deep Learning Techniques

Selection of deep learning techniques is a thoughtful selection which is aligned with the nature of problem and the input data structure, we combine the model CNN and LSTM to extract local features for justification of sequential data which often contains local pattern as periodic signal detecting the CNN excel for key operation like convolution, feature extraction, pooling, from neighboring time steps. Purpose of using LSTM is to capture sequential dependencies and make a justification of model sequences where LSTMs are designed for past information impacts current predictions. It has an ability to handle long term dependencies via different gates and the memory cells. Merging both techniques is helpful to capture both local patterns and sequential relationship. Local pattern via CNN and sequential relationship via LSTM.

Incorporating CNN for local spatial patterns with LSTMs for long term temporal relationship, the design of the model is explicitly built to handle heterogenous data that is sequential. While reducing overfitting and normalizing training consecutive

Conv1D layers with Max Pooling and Normalization by Batch to extract topological features. LSTM layers aggregate correlations into a feature vector in the second layer and process time series in the first layer generating cycles for further analysis. A dropout employed to enhance generalization which allows dense layer to fully understand all of the data from CNN and LSTM outputs although a sigmoid activation in the output layer turns prediction to likelihood for binary classification, ReLU is employed for activation in layers that are hidden for more rapid convergence. Give a little learning rate like as 0.00001, the optimizer developed by Adam assures accurate weight adjustment for exact optimization, binary cross-entropy lack reduces errors in prediction. Classifying like batch normalization and dropout helps this hybrid strategy to successfully identify both short-term and long-terms behaviors hence assuring scalability efficiency fit for consecutive analysis of data.

## *3.4 Hybridization Strategy*

The hybridization strategy that was done in this model involves combining CNN and LSTM to enhance their individual ability in processing sequential data. This method effectively extracts spatial and local patterns which includes the short-term dependencies and correlations among neighbouring features using CNN layers. The Successive convolutional network together with ReLU activation have enabled the creation of layered feature representation, enabling the movement from simple to more complex patterns. Batch Normalization layers assist the stabilization and normalization of activation functions, while the Maxpooling layer lowers the number of dimensions of the feature maps and the long-term temporal dependencies that was present in sequential data are then captured by the LSTM layers. The overall process involves the input sequence to process the first LSTM layer while maintaining its sequential structure followed by the second LSTM layer which summarizes the learned temporal dependencies in a single feature vector. By using concatenation of CNN and LSTMs layer outputs, the model can thoroughly represent both spatial and temporal information effectively.

## *3.5 Training the Model*

The process of training the model was thoughtfully planned to guarantee the convergence and avoid overfitting. The first step of the model training process is to split the dataset into training and testing sets with a ratio of 80:20, this split will ensure that unseen data for final evaluation can still be available. In further enhancing of the learning process, we standardized the features using the Standardize scaler, this will help in normalizing the data and will ensure that the gradients update during the normalization will be stable. After the previous step which involves the preprocessing of data, the pre-processed data is then reshaped into a three-dimensional format so that it will meet the requirements of the hybrid CNN-LSTM architecture in its input layer. To assure that the model is ideal for binary classification task, it was compiled using the Adam optimizer together with a small learning rate of 0.0001 for fine-grained updates and binary cross-entropy for the loss function. A 20% validation split is then utilized throughout the training phase to track the model's generalization performance. Additionally, an Early Stopping callback is implemented to stop training if the validation loss hasn't improved for five consecutive epochs. Together with this combined approach, the mitigation of overfitting can be done effectively and will also ensure the robust performance of the model.

### *3.6 Validation and Testing Model*

The methods used in validating and testing the model is conducted with the use of both data validation conducted in the training and the separation of the post-training datasets to verify the generalization ability of the model. The validation of data has provided immediate feedback on how well the model was performing during its training process, this helps ensure that there is no overfitting in the training set. The completion of training process then marks the start of the model's evaluation based on its effectiveness against the unseen data using the test dataset, this will help in providing the final measurement of the model's accuracy and loss. The reliability on the prediction of the model depends on using a maintained consistent preprocessing method for both training and data testing. The result of our model's evaluation has highlighted its excellent performance obtaining 99.999% accuracy and binary cross-entropy loss of only 0.0002. This near-perfect accuracy of the model is a strong statement that our model can effectively learn the pattern of the data together with its ability to maintain robust generalization. The utilization of the validation and testing methods can help ensure that the indicated performance is not only based on the reflection of the training data but will also reflect to the model's ability to handle unseen scenarios in real-world.

## 4 Experimental Results and Discussion

This section presents the results of the proposed method, its precision and accuracy and discusses what these results entails. The results are also compared with results from previous research.

## 4.1 Results and Discussion

Figure 3 shows the remarkable performance of our proposed model together with its fast learning and high generalizing ability. In the first epoch, the training accuracy have risen significantly and in the third epoch the accuracy has reached on almost 100%, which was then stabilized after that. Through the seen maintainable comparable levels all throughout, it highlights the model's capacity to perform excellently on unknown data, which was also supported by the validation accuracy that closely follows the training curve. The tight alignment of the training and validation accuracy shows that the model is well-balanced, and it preserves a strong generalization while efficiently catching to the trends in the dataset. With great precision and extraordinary efficiency, the consistent performance over the epochs suggests to a well-designed model that is successful for the problem. This finding highlights the validity of the model and shows how ready it is for practical applications.

Figure 4 shows the training and validation loss, over the eight epochs, the graph indicates a notable drop in both training and validation loss, which in the last epoch losses its stability and reached on almost zero. Since it regularly reduces the mistakes over both seen and unseen data, the two curves alignment have suggested a good generalization and effective learning of our proposed model. This presents that our model is strong and well-trained fitting for useful purposes.

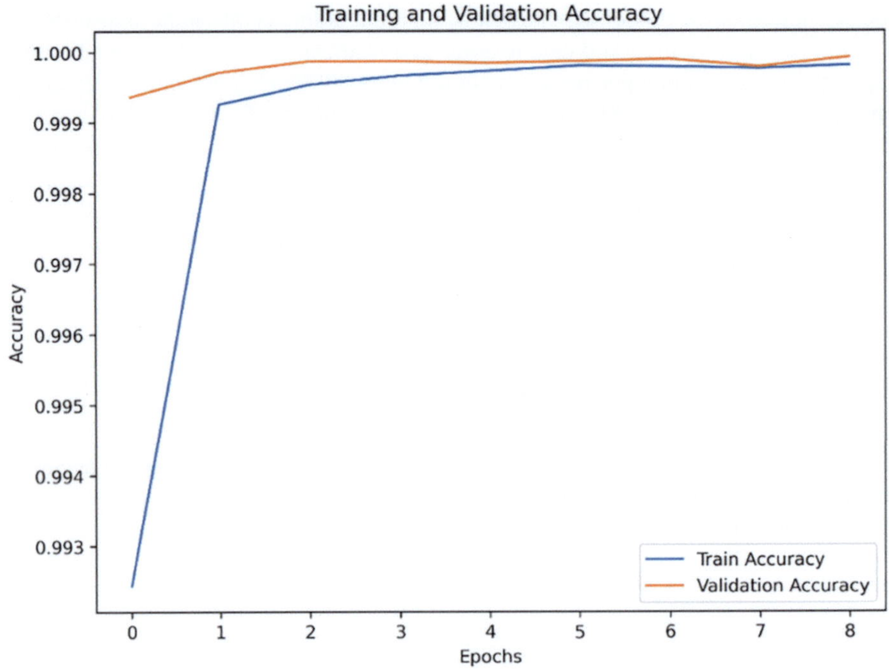


**Fig. 3** Training and validation accuracy of CNN-LSTM model

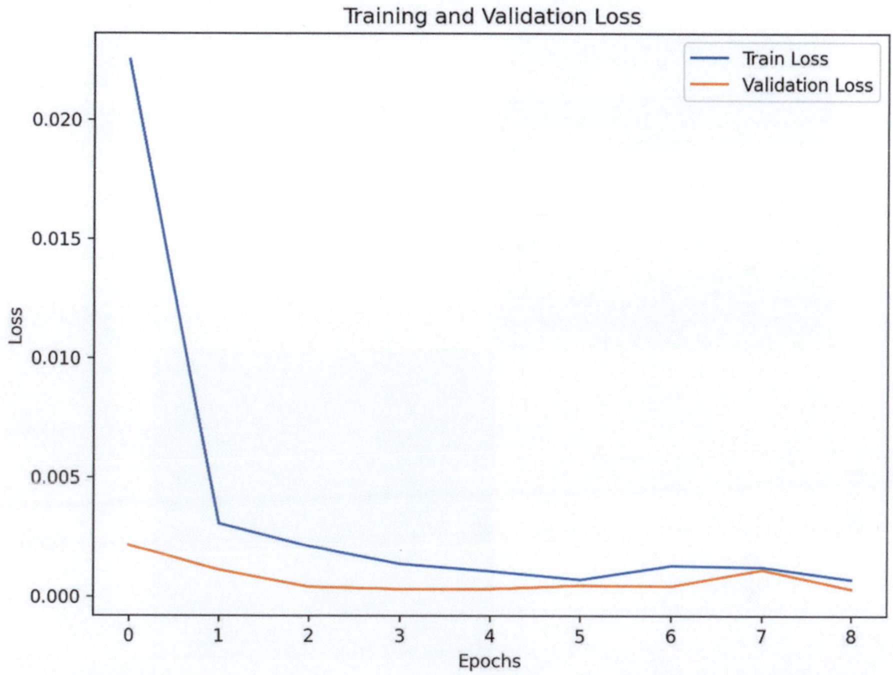


**Fig. 4** Training and validation loss of CNN-LSTM model

### *4.2 Accuracy and Precision*

The classification result and confusion matrix prove the outstanding precision of 99.999% on the research we conducted. With just three incorrect classifications, effectively classified all 27,035 samples with positive result and 20,120 among the 20,124 samples that were negative displaying in Fig. 5, the precision, F1-scores, and recall of 99.999% for both classes shown in Table 1. Achieving 99.999% also overall and average weights of the metrices thus emphasizing the uniformed and harmonious performance of the framework over the two classes, these outcomes reflect the framework consistency, effectiveness in precisely categorizing the data and its superiority compared to other hybrid deep learning approaches which is shown in Table 2.

## 5 Conclusion and Future Work

The result of the study emphasizes the superiority of deep learning over traditional techniques and shows how well it can classify phishing URLs. The model presented in this paper demonstrates a notable increase in accuracy when detecting phishing

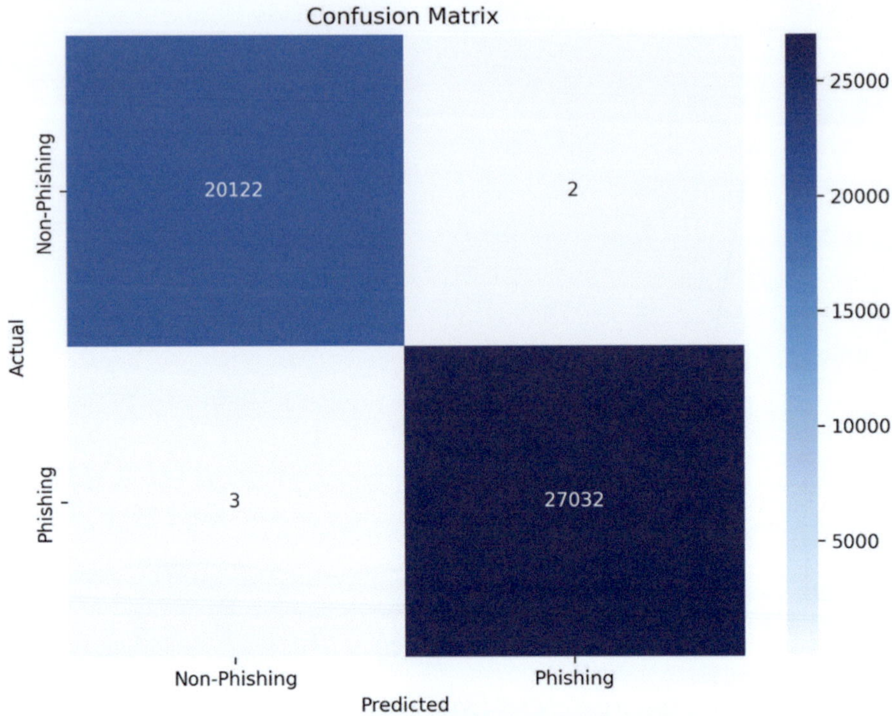


**Fig. 5** CNN-LSTM model confusion matrix

**Table 1** CNN-LSTM model performance result table

| Classification report | Precision | Recall | F1-score | Support |
|---|---|---|---|---|
| 0 | 0.99 | 0.99 | 0.99 | 20,124 |
| 1 | 0.99 | 0.99 | 0.99 | 27,035 |
| Accuracy | | | 0.99 | 47,159 |
| Macro Avg | 0.99 | 0.99 | 0.99 | 47,159 |
| Weight Avg | 0.99 | 0.99 | 0.99 | 47,159 |

attempts through the seamless integration of many deep learning approaches. By integrating CNN and LSTM models, we are able to extract local features for the analysis of sequential data, which frequently exhibits local patterns identical to periodic signals. Our model is thus able to dynamically learn from the evolving phishing tactics and can execute an adaptive learning detection mechanism. As discussed in previous section, the proposed model outperforms previous simulated deep learning methods, achieving an ideal level of precision of 99.99%. What makes this model innovative is the advanced extraction of intrinsic features from the URL that can

**Table 2** Comparison table existing hybrid deep learning techniques

| References | Method | Dataset | Accuracy (%) |
|---|---|---|---|
| [21] | BBOA, CNN | Kaggle | 93 |
| [22] | CWGAN, BGGOA, PDSMV3-DCRNN | ISCX_URL 2016<br>Phishing storm<br>Mendeley 2020 | 99.21 |
| [20] | GRU, RNT | Kaggle | 98 |
| [17] | E-LSTM | GCJ, mailing dataset | 95 |
| [14] | ICSOA-DLPEC | Enron Email dataset | 99.72 |
| [9] | DCCNN-SMO | GCJ, Leopard mobile database | 98.55 |
| [23] | CSO-DCNN | Kaggle | 91 |
| Our approach | CNN-LSTM | Kaggle | 99.999 |

reveal character-level relationships inherent in each specific URL, which drastically affects the classifier's performance.

In the future, we aim to enhance our proposed model by integrating adversarial learning or reinforcement learning to address the continuous evolution of phishing attacks. The exploration of adversarial learning can effectively improve the resilience of our model against adversarial attacks such as manipulated URLs as attackers have become smart in this modern era. The other one which is reinforcement learning (RL) can also be explored to enhance the adaptability of our model as it is a good choice for our model to continuously learn from phishing attempts through adjusting the detection strategy without the need of retraining. With the integration of these advanced techniques in the future, our model's performance can be improved and will significantly aid in the mitigation of rising phishing attacks.

**Acknowledgements** This research work is supported by National Science and Technology Council (NSTC), Taiwan Grant No. NSTC112-2221-E-468-008-MY3.

## References

1. Krishna, S., Natarajan, R., Flammini, F., Alfurhood, B.S., Janhavi, V., Gupta, S.K.: Web security in the digital age: artificial intelligence solution for malicious website classification. IJSWIS **21**(1), 1–25 (2025). https://doi.org/10.4018/IJSWIS.369823
2. Ishtaiwi, et al.: Next-gen phishing defense enhancing detection with machine learning and expert whitelisting/blacklisting. IJCAC **14**(1), 1–17 (2024). https://doi.org/10.4018/IJCAC.353301
3. Hijji, M., Alam, G.: A multivocal literature review on growing social engineering based cyber-attacks/threats during the COVID-19 pandemic: challenges and prospective solutions. IEEE Access **9**, 7152–7169 (2021). https://doi.org/10.1109/ACCESS.2020.3048839
4. Jampen, D., Gür, G., Sutter, T., Tellenbach, B.: Don't click: towards an effective anti-phishing training. A comparative literature review. Human-centric Comput. Inf. Sci. **10**(1), 33. https://doi.org/10.1186/s13673-020-00237-7

5. Gupta, B.B., Arachchilage, N., Psannis, K.: Defending against phishing attacks: taxonomy of methods, current issues and future directions. Telecommun. Syst. **67** (2018). https://doi.org/10.1007/s11235-017-0334-z
6. Vajrobol, V., Gupta, B.B., Gaurav, A.: Mutual information based logistic regression for phishing URL detection. Cyber Secur. Appl. **2**, 100044 (2024). https://doi.org/10.1016/j.csa.2024.100044
7. Zhou, L., Gaurav, A., Alhalabi, W.A., Arya, V., Alharbi, E.: Integrating AI and semantic web technologies for robust phishing detection in virtual realities. IJSWIS **21**(1), 1–19 (2025). https://doi.org/10.4018/IJSWIS.371415
8. Sarker, H., Janicke, H., Ferrag, M.A., Abuadbba, A.: Multi-aspect rule-based AI: methods, taxonomy, challenges and directions towards automation, intelligence and transparent cybersecurity modeling for critical infrastructures. Internet of Things **25**, 101110 (2024). https://doi.org/10.1016/j.iot.2024.101110
9. Said, Y., Alsheikhy, A.A., Lahza, H., Shawly, T.: Detecting phishing websites through improving convolutional neural networks with Self-Attention mechanism. Ain Shams Eng. J. **15**(4), 102643 (2024). https://doi.org/10.1016/j.asej.2024.102643
10. Bahaghighat, M., Ghasemi, M., Ozen, F.: A high-accuracy phishing website detection method based on machine learning. J. Inf. Secur. Appl. **77**, 103553 (2023). https://doi.org/10.1016/j.jisa.2023.103553
11. Al-Subaiey, Al-Thani, M., Abdullah Alam, N., Antora, K.F., Khandakar, A., Uz Zaman, S.A.: Novel interpretable and robust web-based AI platform for phishing email detection. Comput. Electr. Eng. **120**, 109625 (2024). https://doi.org/10.1016/j.compeleceng.2024.109625
12. Nishant, S.P.R., Baitha, S., Kumar, K.D.: An ensemble classification model for phishing mail detection. Proc. Comput. Sci. **233**, 970–978 (2024). https://doi.org/10.1016/j.procs.2024.03.286
13. Bountakas, P., Xenakis, C.: HELPHED: hybrid ensemble learning phishing email detection. J. Netw. Comput. Appl. **210**, 103545 (2023). https://doi.org/10.1016/j.jnca.2022.103545
14. Brindha, R., Nandagopal, S., Azath, H., Sathana, V., Joshi, G.P., Kim, S.W.: Intelligent deep learning based cybersecurity phishing email detection and classification. CMC **74**(3), 5901–5914 (2022). https://doi.org/10.32604/cmc.2023.030784
15. Biswas, A.M., Kumar, A., Delen, D.: A hybrid framework using explainable AI (XAI) in cyber-risk management for defence and recovery against phishing attacks. Decision Support Syst. **177**, 114102 (2024). https://doi.org/10.1016/j.dss.2023.114102
16. Mughaid, S., AlZu'bi, A., Hnaif, S., Taamneh, A., Alnajjar, Elsoud, E.A.: An intelligent cyber security phishing detection system using deep learning techniques. Cluster Comput. **25**(6), 3819–3828 (2022). https://doi.org/10.1007/s10586-022-03604-4
17. Markkandeyan, S., Ananth, A.D., Rajakumaran, M., Gokila, R.G., Venkatesan, R., Lakshmi, B.: Novel hybrid deep learning based cyber security threat detection model with optimization algorithm. Cyber Secur. Appl. **3**, 100075 (2025). https://doi.org/10.1016/j.csa.2024.100075
18. Jain, K., Gupta, B.B., Kaur, K., Bhutani, P., Alhalabi, W., Almomani, A.: A content and URL analysis-based efficient approach to detect smishing SMS in intelligent systems. Int. J. Intell. Syst. **37**(12), 11117–11141 (2022). https://doi.org/10.1002/int.23035
19. Bashir, S., Maqbool, F., Khan, F.H., Abid, A.S.: Hybrid machine learning model for malware analysis in android apps. Pervasive Mob. Comput. **97**, 101859 (2024). https://doi.org/10.1016/j.pmcj.2023.101859
20. Alsubaei, F.S., Almazroi, A.A., Ayub, N.: Enhancing phishing detection: a novel hybrid deep learning framework for cybercrime forensics. IEEE Access **12**, 8373–8389 (2024). https://doi.org/10.1109/ACCESS.2024.3351946
21. Gupta, B., et al.: A hybrid CNN-brown-bear optimization framework for enhanced detection of URL phishing attacks. CMC **81**(3), 4853–4874 (2024). https://doi.org/10.32604/cmc.2024.057138

22. Prasad, Y.B., Dondeti, V.: PDSMV3-DCRNN: a novel ensemble deep learning framework for enhancing phishing detection and URL extraction. Comput. Secur. **148**, 104123 (2025). https://doi.org/10.1016/j.cose.2024.104123
23. Gupta, B.B., et al.: Cuckoo search-optimized deep CNN for enhanced cyber security in IoT networks. CMC **81**(3), 4109–4124 (2024). https://doi.org/10.32604/cmc.2024.056476

# Security Implications of Quantum-Resistant Encryption in Emerging IoT Networks

**Eric Blancaflor, Mark Francis James Bona, David Josiah Dychioco, Matthew Jake Inson, and Prince Rayly Reyes**

**Abstract** With Industries already feeling the impact of IoT devices and more especially Gadgets, the rise of quantum computing that is set to happen in the next foreseeable future presents a major challenge to most of the encryption techniques implemented in the industries to secure such networks. A lot of work has been done in recent years regarding quantum-resistant cryptography as a broad topic, but less is known about how it applies to IoT networks that are particularly bounded in terms of processing power and have heterogeneously structured messaging channels. This research will seek to evaluate the issues connected with a quantum-threatened future and the possibility of genuinely integrating post-quantum cryptography into IoT systems. It will consider how they affect the performance of the device, the rate of network latency and the overall security outlook to give a detailed assessment of how these new generation encryption methods can be harnessed to protect the rapidly expanding IoT systems from hack threats in the quantum environment.

**Keywords** Internet of things (IOT) · Security · Quantum-resistant encryption · Post-quantum encryption · Public key encryption with keyword search · Inside keyword guessing attacks · Public key authenticated encryption with keyword search · Quantum computing vulnerable

## 1 Introduction

### *1.1 Background of the Study*

There has been the growth in the Internet of Things (IoT) technologies in the current society especially in smart homes, industries and healthcare. However, this expansion has new risks or cyber threats especially at the edges where IoT devices are sited. The

E. Blancaflor (✉) · M. F. J. Bona · D. J. Dychioco · M. J. Inson · P. R. Reyes
School of Information Technology, Mapua University, Manila, Philippines
e-mail: ebblancaflor@mapua.edu.ph

 
Y. Ma (ed.), *Proceedings of 4th International Conference on Intelligent Systems Design and Engineering Applications*, Lecture Notes in Electrical Engineering 1478,
https://doi.org/10.1007/978-981-95-1802-9_4

IoT infrastructures are somewhat entirely different from what the conventional security solutions were created to protect and thus the current approaches are not adequate to meet the emerging IoT requirements. Since it is estimated that in 2025 there will be 30.9 billion IoT devices, strong security is imperative. While there are ideas about dynamic security policies, most of these are very concrete and tailored to specific issues; and thus, not very adaptive for the current dynamics of the IoT. In addition, they may not contain important elements of design including risk management, legal issues, or AI-based dynamic adjustability mechanism. The SolarWids attack is one of the examples of recent cyberattacks indicating the need for a new paradigm in protecting IoT systems from the modern threat [1]. Furthermore, they often do not include key components such as risk management plans, legal concerns, or AI-based dynamic adjustability strategies. The SolarWinds attack is just one example of recent cyberattacks that highlight the necessity for new approaches to IoT system protection against modern threats [2].

As cloud-computing is progressing, many organizations benefit from cloud storage technologies to minimize their expenditures. However, for security purposes, data should be encrypted before being stored in cloud networks and storage spaces. Public Key Encryption with Keyword Search (PEKS), proposed by Boneh et al., allows for flexible use of encrypted data. Regrettably, the majority of the proposed PEKS schemes are insecure against inside keyword guessing attacks (IKGA), where attackers may reveal the keyword information in the trapdoor. To counter this, Huang and Li proposed Public Key Authenticated Encryption with Keyword Search (PAEKS), in which the trapdoor of the receiver is constructed such that it is only valid for authenticated ciphertexts. Following their work, many PAEKS schemes have been proposed to improve security, with some considering potential future quantum attacks. Quantum computers are expected to break traditional encryption methods like RSA and ECC, prompting the need for quantum-resistant cryptography. Lattice-based cryptography, hash-based signatures, and other post-quantum algorithms are being explored to mitigate this risk. However, our cryptanalysis reveals that many of these PAEKS schemes remain vulnerable to IKGA attacks, even in a quantum-resistant context. This emphasizes the need for ongoing research to address these vulnerabilities [3].

Due to the cross-cutting nature of securing IoT systems, design, technology, people and tangible objects are involved in multiple ways. Establishing a robust security system requires a holistic approach that synchronizes both hardware and software functions. This synchronization is crucial for building foundational security layers within operating systems to safeguard computing systems across various platforms and conditions. As IoT continues to grow rapidly, so do the security threats. Hence, it is crucial to design extensive security measures, including insurance against quantum coup—a feature of present and future threats to IoT systems. Such topics of research must be oriented toward the integration of quantum-resistant algorithms into IoT devices and security-based optimization of the devices' scarce computational capabilities [2].

## 1.2 Significance of the Study

This general contribution of this paper is that it goes beyond addressing practical technical concerns while drawing attention to some possible dire consequences of quantum computing within the confines of the IoT environment. Future researchers may be motivated, thru this study, to look on the part of cryptographers, producers of the devices and policymakers in order to effectively counter the increasing attack prevalence of IoT devices, it becomes increasingly necessary for security architectures to develop commensurately to prevent attacks on personal data and to sustain public confidence in emerging technologies. By emphasizing a clear and practical analysis of post-quantum cryptographic solutions, the research aims to contribute significantly to the ongoing discourse on securing the future of IoT technologies against emerging quantum threats.

## 1.3 Research Gap

Despite the alarming implications of quantum attacks, there remains a significant knowledge gap regarding the integration of quantum protection measures within the IoT domain. The theoretical construction of post-quantum cryptography (PQC) may be a hot topic, but actual analysis regarding the integration of these algorithms using conventional IoT devices is rather limited. If we look at IoT as a set of connected devices, the latter are often bound by restricted resources such as low processing power, limited available energy, and the varying degrees of network integration. Consequently, implementing sophisticated quantum-resistant algorithms in such environments poses unique challenges. It is essential to explore how these algorithms can be adapted or optimized for devices that inherently lack the resources to support traditional cryptographic frameworks [4].

Moreover, existing research tends to focus on the theoretical advancements of quantum-resistant algorithms, frequently overlooking the practical implications for IoT infrastructure. For example, the implications of implementing quantum-resistant encryption to traffic density of networks, the ability of the devices, and scalability of the whole system remain unknown. This lack of research direction is an area of study that offers significant potential for exploration to elucidate how QR-cryptography should be deployed in IoT systems, particularly where constrained computational resources are a consideration. When considering these issues, the authors can help to improve the stability and the protection of the IoT environment [5].

In addition, there is a notable absence of studies addressing the evolution of IoT security protocols towards incorporating quantum-resistant encryption. Many existing IoT devices currently operate under security frameworks that may not support the advanced encryption standards required to mitigate quantum computing threats. It also defined a research gap—a prime area of improvement in the current literature on the technical and structural considerations for deploying quantum-resistant cryptography in IoT networks. Thus, by narrowing the field of research to the combination of quantum resilience and IoT security, investigators are able to open up new horizons in defining novel ways of improving the security and functionality of IoT systems as businesses move into a quantum era [6].

## 1.4 Objective of the Study

The general objective of this research is to assess the security concerns associated with implementing quantum-resistant cryptography in next-generation IoT networks, as quantum computing poses an increasing threat to traditional cryptographic methods. Specific objectives are:

- Examine specific vulnerabilities in IoT networks that could be exploited by quantum-powered attacks and assess the risks tied to the continued use of classical encryption in these networks.
- Compare and evaluate quantum-resistant encryption schemes for their effectiveness in securing IoT devices, while considering resource constraints like processing power and energy consumption.
- Investigate the practical challenges of deploying quantum-resistant encryption in IoT, such as potential increases in computational load, network delay, and energy requirements, to inform a long-term strategy against quantum threats.

## 1.5 Scope and Delimitation

The work of this paper is to examine the security concerns when implementing quantum-resistant encryption (QRE) in new-era Internet of Things (IoT) networks. That is why, considering the further development of the IoT devices distribution throughout all sectors, the security question has become essential. When these devices interact and exchange messages, their capabilities to perform computations and store enormous data hamper the practical aspect of cryptographic systems. This study will focus on how practical post-quantum cryptographic algorithms can be embedded into low-power IoT devices while addressing their security vulnerabilities [7]. Researcher have underlined that it is crucial to adjust cryptographic measures used for protecting these devices to new threats, especially those that originate from quantum computing, which poses a threat to most of the current cryptographic methods commonly used, such as RSA and ECC [8]. Furthermore, with reference to post-quantum cryptographic algorithm, the research seeks to evaluate the extent of the use of some distinct algorithms such as lattice based, hash based and multivariate polynomial. Existing studies have shown that while some post-quantum algorithms offer promise, their efficiency and practicality within resource constrained IoT environments remain under-explored [9]. This work will not seek to develop new quantum-resistant algorithms but will assess the security threats in the utilization of current algorithms in IoT forms of networks and systems. Studying the modern cryptographic systems, the research will reveal their advantages and disadvantages as protective tools against possible threats that may come from quantum attackers and shielding the IoT environment as the world extends its presence in people's lives. However, beyond analyzing these purely cryptographic features, the research will also consider performance parameters such as device energy requirements and the network density of a given environment. Given that many IoT applications are expected to involve real-time data processing, optimal security cannot be achieved without regard for operational performance. As has been found earlier in literature, the integration of quantum-resistant algorithms can contribute to raising the computational load, which can harm the device efficiency and durability [10]. Therefore, understanding the trade-offs between security enhancements and practical device constraints will be a central theme of this investigation, providing valuable insights for developers and policymakers. Lastly, while the study will focus on technical and practical considerations, it will not delve into the policy or regulatory implications of implementing quantum-resistant encryption in IoT networks. This technical approach allows for a more concentrated evaluation of the algorithms' suitability without the potential distractions of legal frameworks and compliance issues [11].

## 2 Literature Review

### *2.1 Identify IoT Security Vulnerabilities*

Internet of Things (IoT) is contemporary technology that makes it possible to interface numerous objects present in both physical and virtual worlds. First, the Internet developed from a simple connection between personal computers, to client–server architectures, the WWW, e-mail and file-sharing. What used to be a small network of connections has grown into a gargantuan interconnected body, linking billions of smart devices into advanced systems today. These systems use sensors and actors capable of observing, regulating, and communicating with the physical contexts in which they find themselves [12].

As broadly defined, vulnerabilities are weaknesses in a system that allow one or more individuals accepted from the open community to perform unauthorized activity; in relation to IoT, these may relate to a breakdown of integrity, confidentiality, or availability of services. For this reason, it is vital to have awareness of these vulnerabilities so as to be in a position to start coming up with merits of defense mechanisms. Several works bring a clear picture of IoT vulnerabilities, outlining how they impact on security models. OWASP has defined some of the major risks which are common in IoT devices and are known as the ten most critical IoT risks. There's significant concern on device security since IoT related products are often physically impaired because of hostile tokens; vandals or other forms of unauthorized intruders have easy access to the devices in question since they are bound to be exposed to environmental pressures as well as different forms of adversity. This could result in malicious firmware cloning and data manipulation including cloning of radio frequency signals to unlock electric cars or manipulating a vehicle's controller and area network. Another danger relies on the insecure booting that enables the placement of malicious activities uncontrolled when booting the device; by using nest thermostat and nikel + FuelBand devices, this was illustrated. Network vulnerabilities are also very dangerous which attack connected devices through weak services, improper authentication, low encryption standard & default/newly created passwords. Such weaknesses include DDoS, where multiple legitimate users are overwhelmed by fake ones, Sybil, where one client pretends to be many, or data theft. Small memory and restricted resources in IoT devices significantly impact the enhancements in the encryption, as well as exposes IoT devices to threats; for instance, the medical sector where attackers can seize complete control of devices inclusive of an insulin pump or pacemaker. There is lack of definite cryptanalogue, vulnerable open ports, and low encryption put Iot devices at risk of intrusions including the mirai botnet intrusions that targeted open telnet ports and standard factory login details. SHODAN, Masscan, NMAP are used in order to find out the devices which are prone to open ports. Table 1 shows major IoT security challenges and vulnerabilities [13].

The current IoT ecosystem is riddled with many problems because all the devices lack standard manners to follow to implement security measures for their systems and actors. Inherent device differences and relatively low computation capabilities of many devices render conventional security measures ineffective and non-applicable since cost, power, and size considerations exacerbate the situation. Another issue is that of interoperability as with the increasing number of devices connected, there are [no common platforms]. This is a challenge because a lot of these IoT devices require connectivity to be critical and since connectivity varies between devices this creates a problem. Compared to an environment that is fixed, for instance, connectivity,

**Table 1** Major IoT security challenges and vulnerabilities [13]

| Challenge | Description | Example |
|---|---|---|
| Outdated firmware | Many IoT devices run on outdated, vulnerable firmware | Smart TVs, household appliances |
| Weak passwords | Devices often come with default or weak passwords | Consumer IoT devices |
| Network vulnerabilities | Weak encryption and authentication methods lead to attacks like DDoS, data theft | Sybil attacks, botnets like Mirai |
| Insufficient computation power | Limited processing power and memory restrict advanced security measures | Medical devices, wearables |
| Insecure booting | Vulnerabilities in the boot process allow malicious software to be introduced | Smart thermostats, vehicle controllers |

especially in cars, raises other security issues. Lack of standardization from the government means that the service providers can come up with different solutions for the problem thus failing to create a communal IoT platform. Security continues to be one of the fundamental aspects that would enable consumers to use the products however, due to costs and other constraints such as size and power there is always ys compromised in the kind of security that is offered to enable the IoT systems hence keep on being at risk of being attack with severe consequences including very huge losses and losses of reputation. While expectations for IoT are increasing because of AI improvement and data engineering, these unsolved questions enhance the reluctance of actors. Previous papers have partially discussed these challenges where several surveys have compared IoT security, DDoS attacks, IDS, and machine learning techniques for IDS in detail. In Table 2 there is a comparative analysis, and a summary of contributions derived from highly cited research works highlighted in this study area [14].

The integration of quantum-resistant encryption (QRE) into IoT is challenging due to the inherent constraints of IoT devices, such as limited memory, processing power, and energy. Lattice-based algorithms like Learning With Errors (LWE) and Ring-LWE, while robust against quantum attacks, are computationally intensive and require processing capabilities often unavailable in IoT devices. This creates tension between security needs and hardware limitations, especially as real-time IoT applications are highly sensitive to latency [15, 16]. Despite advances in standardizing QRE algorithms, such as by the National Institute of Standards and Technology (NIST), challenges persist—particularly with large key sizes in some quantum-resistant cryptosystems, which can overwhelm IoT devices with limited storage capacity. Addressing this, researchers are developing lightweight versions of QRE that maintain performance and can be deployed across various IoT platforms [17, 18]. Hybrid cryptographic approaches that combine traditional and quantum-resistant algorithms could provide transitional security until quantum computing becomes mainstream, balancing immediate performance with future-proofing needs. Hardware accelerators for post-quantum algorithms may further improve QRE viability in constrained environments, and coordinated efforts among policy makers, industry leaders, and academia are essential to support QRE deployment through incentives

**Table 2** Comparison between different schemes [28]

| Cryptosystem | Size of ciphertext | Access policy | Multi-use | Standard model | Re-encryption verifiability |
|---|---|---|---|---|---|
| DSD21 | 1 + m | ✗ | ✗ | ✓ | ✗ |
| LQZ21 | 1 + m | ✗ | ✓ | ✓ | ✗ |
| SRA20 | 1 + m | ✗ | ✗ | ✗ | ✗ |
| LMZ19 | 1 + 2 lm | AND-gate | ✗ | ✓ | ✗ |
| WYZ21 | 1 + 2 m | ✗ | ✓ | ✓ | ✓ |
| KPAB-VPRE | 1 + 2 lm | Any monotonic | ✓ | ✓ | ✓ |

and regulatory frameworks. Such advancements are crucial, especially in sectors like healthcare, finance, and transportation, where IoT plays a key role [17].

## 2.2 Access Quantum-Resistant Encryption

Quantum-resistant cryptography (QRC) or was established as post-quantum cryptography is a concept that focuses on the algorithms that remain safe from the power of quantum computers, which can perform a million times faster than most advanced technologies in the current world [19]. Although public key cryptosystems including RSA and elliptic curve cryptography (ECC) are secure to date, they are built on problems amenable to efficient quantum solutions and are consequently insecure. On the other hand, QRC draws from mathematical problems which are thought to be unbreakable providing better security to future uses [20–22]. Several categories of quantum resistant algorithms are available in the following: Lattice-based, code-based, hash-based, and multivariate-quadratic cryptography which a quantum computer will take a long time to decode or to decrypt. For example, the normally adopted lattice-based cryptography offers advantage in scalability and, aptly suitable for adoption on resource-restricted devices such IoT networks, thereby extends its relevance in gadgetry like mobile phones and smart home appliances which do not strictly require sophisticated cryptography. These devices which form the core of smart home networks are a potential threat as quantum computing advances, new avenues of data risks to consumers emerge. Post-quantum lattice-based solutions apply Learning with Errors (LWE) and Ring Learning with Errors (RLWE) modes which provide both encryption and digital signatures and are generally both efficient and resistant to all potential attacks from quantum algorithms [10, 23, 24].

Another two notable algorithms are Merkle tree-based, which is hash-based cryptography, and McEliece encryption, which is code-based cryptography. Both are proven to be secure and quantum-resistant, but it has a similar weakness with RSA and EEC, which relies on large key sizes to be secure [25–27]. On the Internet of Things, it is quite understandable that these encryptions are not applicable as the hardware of these things are not made with encryption in mind. They are powerful enough to serve their purpose, but not powerful enough to encrypt everything that it does. This is one of the main reasons why these algorithms, although secure and quantum-resistant, are not applied to IoT without further research.

Moreover, the U.S National Institute for Standard and Technology (NIST) has been making a continuous effort to standardize quantum-resistant cryptography algorithms to ensure security and safer cyber environments. Some of these algorithms have passed and are currently being used in testing as of 2022. This is significant as these algorithms can serve as the baseline of IoT security and quantum-resistant infrastructure that can significantly contribute to private sectors such as finance, healthcare, or other sectors that require critical security [20, 22]. Furthermore, the transition to quantum-resistant systems is essential to ensure long-term security and to mitigate potential risks associated with the advent of quantum computing.

Table 2 illustrates a comparative analysis of our KPAB-VPRE scheme with other relevant schemes across various attributes, including ciphertext size, access policy expression, multi-use functionality, security model, and re-encryption verifiability. As demonstrated in Table 2, some schemes lack support for access policy expression entirely, while certain others, such as a specific scheme, only support "AND" expressions. In contrast, our scheme utilizes the LSSS matrix to articulate access policies, allowing for more complex operations such as "AND," "OR," and "THRESHOLD." Moreover, matrix operations are applied to the encryption and decryption algorithms, which contributes to improved efficiency. In terms of re-encryption verifiability, only the scheme proposed by Wu et al. includes this feature, though it does not support flexible access policies. Thus, the proposed KPAB-VPRE scheme offers fine-grained control and secure sharing of encrypted data while also supporting multi-use functionality and re-encryption verification, making it a highly practical solution [28]. Integrating quantum-resistant encryption methods could further enhance the robustness of this scheme, safeguarding it against the computational capabilities of quantum technologies. Implementing such quantum-safe encryption would reinforce data integrity and secure access policies, ensuring resilience in a post-quantum environment.

## 2.3 *Examine Real-World Implementation Challenges*

Quantum computers can accomplish things our computers can that are magnitudes faster, and what supercomputers can't, and because of this, it could jeopardize widely used encryption algorithms such as RSA and ECC, the basic building blocks of modern internet security methods. Firstly, Shor's algorithm exponentially factorizes large integers and can solve discrete logarithm problems exponentially faster than modern computers, something that widespread cryptosystems cannot defend against and insecure [29]. The implications of these are vast, from securing digital communications up to the protection of sensitive information, and the establishment of trusts in online transactions, these are some of the things that should be protected, hence security controls are needed to be established, which includes the defense from quantum computing attacks.

Current encryption standards such as RSA rely on the difficulty of factoring large prime numbers and ECC is based on the discrete logarithm problem on an elliptic curve. These vulnerabilities could be exploited by quantum computers, undermining the security assumptions that have governed cryptographic practices for decades. With increasingly powerful and accessible quantum computers, these cryptosystems are inefficiently compromised by these computers, and with those would come significant risks if sensitive internet data were to be transmitted [30]. This is particularly concerning for Internet of Things (IoT) networks, where devices often use lightweight implementations of these cryptographic algorithms to conserve resources.

Figure 1 shows the expanding IoT landscape, comprising devices that transmit sensitive data like health information, financial transactions, and critical infrastructure controls, is increasingly vulnerable to quantum-based attacks. Many IoT devices are intentionally designed with limited computational and energy resources, leaving them susceptible to future quantum-enabled threats, as current cryptographic algorithms may not withstand the power of advanced quantum systems. This exposure could lead to breaches where malicious actors intercept communications, manipulate devices, or disrupt entire networks [31]. In sectors like healthcare, finance, and critical infrastructure, such vulnerabilities could enable quantum-equipped attackers to cause significant disruption, identity theft, or financial fraud. With the advent of quantum computing on the horizon, current cryptographic defenses risk obsolescence, highlighting the urgency of deploying quantum-resistant algorithms capable of withstanding quantum threats within the IoT ecosystem [32]. To meet these challenges, the cryptographic community, with organizations such as the U.S. National Institute of Standards and Technology (NIST), has been working to identify and standardize post-quantum algorithms that balance security with feasibility in resource constrained IoT environments [19, 20, 22]. Transitioning to quantum-resistant algorithms is critical not only for protecting today's systems but also for ensuring IoT reliability in sectors where security breaches would have catastrophic impacts, such as autonomous vehicles and industrial automation.

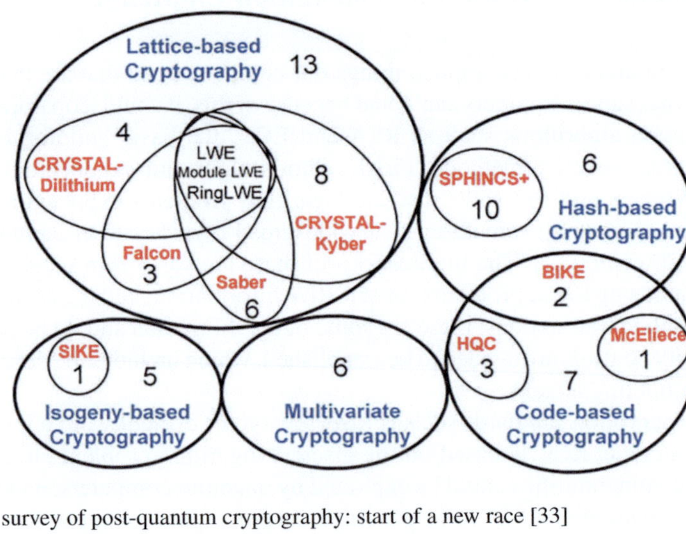


**Fig. 1** A survey of post-quantum cryptography: start of a new race [33]

## 2.4 Synthesis

The development of segment based IoT networks has become one of the most significant trends in modern society and industry, as billions of interconnected devices now monitor, control, and adjust both physical and digital environments. However, this interconnectivity brings new security challenges. Many IoT devices have limited computational power and rely on basic, often insecure cryptographic tools. Traditional encryption methods like RSA and ECC are increasingly vulnerable as quantum computing advances, heightening the risks for IoT applications ranging from personal health data to industrial control systems. These shortcomings clearly explain why it is crucial to respond to the insecurity threats existing in the IoT environments to protect data integrity, confidentiality as well as network availability. The promising method to protect communication against quantum threats is quantum-resistant cryptography (QRC) or post-quantum cryptography. It employs many encryption algorithms that were developed with a capability of overcoming all forms of quantum decryption such as Shor's algorithm. For example, lattice-based cryptography is suitable for IoT devices due to its robustness against quantum attacks and relative efficiency on low-power devices. However, real-world deployment of QRC in IoT systems still faces challenges due to the computational and energy constraints of IoT devices; the large key sizes and processing demands of quantum-resistant algorithms can disrupt normal device operations. While the U.S. National Institute of Standards and Technology (NIST) is working to standardize post-quantum algorithms, these algorithms need further optimization to function effectively and remain compatible with IoT environments. To facilitate the transition, combining widely used symmetric/asymmetric cryptography methods with post-quantum techniques may offer an immediate, adaptable solution. Specific hardware accelerators for QRC could further improve processing speed in constrained devices, helping IoT networks become quantum-resilient and better prepared for the post-quantum era.

# 3 Conclusion

This paper has thus investigated and discussed the security concerns when QRE is deployed within today's IoT network. The number and use of IoT have grown rapidly in the past few years, which have caused disruption and improvement across several industries in terms of productivity and networking but also introduced several risks to security. Due to the general limitation of IoT devices in terms of computing and memory the need for good encryption capable of withstanding the challenges that are posed the quantum computing is highly required. Evaluate current Iteration post-quantum cryptographic algorithms like lattice-based, hash based and multivariate polynomial; we had seen that they have advantages and disadvantages compared to device/community optimality and power utilization. Therefore, the results of this research are indicating the greatest significance of integrating quantum-safe cryptography into IoT systems. Although there may be no need for developing entirely new algorithms at this time, it is compulsory to assess the security threats posed by the current frames. In understanding the analysis of these algorithms, the envisaged interest seeks precise exposition of working overlays that can fathom out ways of improving security whilst not jeopardizing the functionality of IoT devices. Further, our work stresses the importance of reflexive protective mechanisms to respond to the emerging IoT solutions and new challenges linked to quantum computing.

In conclusion, in this paper, it is proposed to deepen the investigation of quantum-resistant approaches for checking the stability of IoT networks in the future. Our work aims at shedding light on the implementation of QRE, regarding both obstacles and possibilities, to help create stronger safety nets for IoT environments. The idea here is that further discussions and cooperation between the above involved parties will be crucial while managing the emerging quantum threats as well as spurring innovation in an environment that is optimally secure.

## References

1. Halgamuge, M.N., Niyato, D.: Adaptive edge security framework for dynamic IOT security policies in diverse environments. Comput. Secur. **148** (2025). https://doi.org/10.1016/j.cose.2024.104128
2. Sun, P., Wan, Y., Wu, Z., Fang, Z.: A survey on security issues in IOT operating systems. J. Netw. Comput. Appl. **231** (2024). https://doi.org/10.1016/j.jnca.2024.103976
3. Liu, Z.-Y., Tseng, Y.-F., Tso, R., Mambo, M., Chen, Y.-C.: 2022. Public-key authenticated encryption with keyword search. In: Proceedings of the 2022 ACM on Asia conference on computer and communications security (May 2022), pp. 423–436. https://doi.org/10.1145/3488932.3497760
4. Ghosh, S., Zaman, M., Joshi, R., Sampalli, S.: Multi-phase quantum resistant framework for secure communication in SCADA systems. IEEE Trans. Dependable Secure Comput. **2024**, 1–18 (2024). https://doi.org/10.1109/tdsc.2024.3378474
5. Kumar, M., Pattnaik, P.: Post quantum cryptography(pqc)—an overview: (invited paper). In: 2020 IEEE High Performance Extreme Computing Conference (HPEC) (2020). https://doi.org/10.1109/hpec43674.2020.9286147
6. Saki, A.A., Alam, M., Phalak, K., Suresh, A., Topaloglu, R.O., Ghosh, S.: A survey and tutorial on security and resilience of quantum computing. In: 2021 IEEE European Test Symposium (ETS) 4 (May 2021), pp. 1–10 (2021). https://doi.org/10.1109/ets50041.2021.9465397
7. Alotaibi, A., Chikh, A.: A survey on quantum-resistant cryptography: challenges and solutions. IEEE Access **9**(2021), 128535–128558 (2021). https://doi.org/10.1109/ACCESS.2021.3111178
8. Chen, L.K., Wang, Y.: Quantum-resistant cryptography: an overview. Futur. Gener. Comput. Syst. **109**(2020), 204–213 (2020). https://doi.org/10.1016/j.future.2020.02.052
9. Güneş, M., Koç, Ç.: Performance evaluation of post-quantum cryptography in IoT applications. Sensors **21**(10), 3495 (2021). https://doi.org/10.3390/s21103495
10. Asif, R.: Post-quantum cryptosystems for internet-of-things: a survey on lattice-based algorithms. IoT 2, 1 (February 2021), pp. 71–91 (2021). https://doi.org/10.3390/iot2010005

11. Thales Group. Post-Quantum Cryptography Solutions. Retrieved October 22, 2024 from https://cpl.thalesgroup.com/encryption/post-quantum-crypto-agility?utm_source=google&utm_medium=paid&utm_campaign=2024q3-ds-apac-asean-pqc-search&utm_content=705425362000&utm_keyword=quantum+proof+cryptography&gclid=Cj0KCQjwmt24BhDPARIsAJFYKk1zIwzUao4YT_othHw6Fbw8S2s11SBqz0BaAYxQWsC8DMG7zXvzLo8aAk5dEALw_wcB&gad_source=1
12. Azrour, M., Mabrouki, J., Guezzaz, A., Kanwal, A.: Internet of things security: challenges and key issues. Secur. Commun. Netw. **11**. (2021) https://doi.org/10.1155/2021/5533843
13. Malhotra, P., Singh, Y., Anand, P., Bangotra, D.K., Singh, P.K., Hong, W.-C.: Internet of things: evolution, concerns and security challenges. Sensors **21**(5), 1809 (2021). https://doi.org/10.3390/s21051809
14. Mishra, N., Pandya, S.: Internet of things applications, security challenges, attacks, intrusion detection, and future visions: a systematic review. IEEE **9**(April 2021), 59353–59377 (2021). https://doi.org/10.1109/access.2021.3073408
15. Wang, X., Xu, G., Yu, Y.: Lattice-based cryptography: a survey (2023). Retrieved from https://doi.org/10.1007/s11401-023-0053-6
16. Espitau, T., Wallet, A., Yu, Y.: On Gaussian sampling, smoothing parameter and application to signatures (2023). Retrieved from https://doi.org/10.1007/978-981-99-8739-9_3
17. Goyal, R.: Quantum cryptography: secure communication beyond classical limits (2024). Retrieved from https://doi.org/10.36676/jqst.v1.i1.01
18. Yesina, M.V., Ostrianska, Y.V., Gorbenko, I.D.: Status report on the third round of the NIST post-quantum cryptography standardization process (2022). Retrieved from https://doi.org/10.30837/rt.2022.3.210.05
19. Mattsson, J.P., Smeets, B., Thormarker, E.: Quantum-resistant cryptography (2021). Retrieved from https://arxiv.org/abs/2112.00399
20. Bavdekar, R., Chopde, E.J., Bhatia, A., Tiwari, K., Daniel, S.J., Atul: Post quantum cryptography: techniques, challenges, standardization, and directions for future research. (2022). Retrieved from https://arxiv.org/abs/2202.02826
21. Kumar, A., Ottaviani, C., Gill, S.S., Buyya, R.: Securing the future internet of things with post-quantum cryptography (2022). Retrieved from https://arxiv.org/abs/2206.10473
22. Xiaoyun Wang, Guangwu Xu, and Yang Yu. 2023. Lattice-Based Cryptography: A Survey. (November 2023). Retrieved from https://link.springer.com/article/https://doi.org/10.1007/s11401-023-0053-6
23. Morsalani, M.: Quantum and post-quantum cryptography: principles, protocols, and challenges (2024). Retrieved from https://doi.org/10.13140/RG.2.2.30121.81767
24. Akter, M.S.: Quantum cryptography for enhanced network security: a comprehensive survey of research, developments, and future directions (2023). Retrieved from https://arxiv.org/abs/2306.09248
25. Weger, V., Gassner, N., Rosenthal, J.: A survey on code-based cryptography (2024). Retrieved from https://arxiv.org/abs/2201.07119
26. Chikouche, N., Ghadbane, A.: Performance evaluation of post-quantum public-key cryptography in smart mobile devices. Springer, Cham. Retrieved from https://doi.org/10.1007/978-3-030-02131-3_9
27. Farooq, M., Khan, R., Hassan Khan, M.: QuantIoT novel quantum resistant cryptographic algorithm for securing IoT devices: challenges and solution (2023). Retrieved from https://doi.org/10.21203/rs.3.rs-3160075/v1
28. Hou, J., Peng, C., Tan, W., Ding, H.: Quantum-resistant multi-feature attribute-based proxy re-encryption scheme for cloud services. Comput. Model. Eng. Sci. **138**(1), 917–938 (2023). https://doi.org/10.32604/cmes.2023.027276
29. Larasati, H.T., Kim, H.: Quantum cryptanalysis landscape of Shor's algorithm for elliptic curve discrete logarithm problem. Springer, Cham. Retrieved from https://doi.org/10.1007/978-3-030-89432-0_8
30. Singh, B., Ahateshaam, Md., Lahiri, A., Sagar, A.K.: Future of Cryptography in the Era of Quantum Computing. Springer Science+Business Media (2024). Retrieved from https://doi.org/10.1007/978-981-99-8661-3_2

31. Alomari, A., Kumar, S.A.P.: Internet of Things (2024). https://doi.org/10.1016/j.iot.2024.101132
32. Şafak, I., Alagöz, F., Anarim, E., Şafak, I., Alagöz, F., Anarim, E.: Post-quantum security measures for the internet of things. (2024). Retrieved from https://doi.org/10.4018/978-1-6684-7366-5.ch075
33. Yadav, S.: An extensive study on lattice-based cryptography and its applications for RLWE-based problems (2023). Retrieved from https://doi.org/10.36676/urr.2023-v10i3-014

# Modern Electronic Systems and Signal Analysis

# FOV-Adaptive Bilateral-Branch Network for Etch Bias Model with SEM Contour

**Feng Chaoxin, Cheng Qian, and Jia Ruoran**

**Abstract** The etch process plays a critical role in CD (Critical Dimension) control for patterning, and as technology nodes shrink, the importance of accurately predicting etch bias has escalated. Recent years have witnessed remarkable advancements in etch bias modeling research, with the emergence of various innovative methods, including machine learning and deep learning, for predicting etch bias. Nevertheless, the current machine learning approach remains an empirical model, relying heavily on the rationality of manually designed features. Deep learning methods, despite their popularity, often rely solely on Convolutional Neural Networks (CNNs) and lack a comprehensive integration of the fundamental principles of the etching process. To address these limitations, we introduce a novel loading effect adaptive etch bias model. This model uniquely determines its input range based on the specific characteristics of the pattern, enabling it better to capture the intricate nuances of the etching process. Compared to previous CNN-based models, our approach demonstrates a superior ability to perceive the loading effect by optimal FOV (Field of Vision), resulting in significantly higher accuracy in etch bias prediction. This innovative model not only enhances the precision of etch bias estimation but also contributes to the overall advancement of patterning technology.

**Keywords** Etch bias model · End-to-end · CNN · Loading effect · Field of vision

---

C. Feng
School of Integrated Circuits, University of Chinese Academy of Sciences, Beijing, China

C. Qian
Institute of Advanced Technology, University of Science and Technology of China, Hefei, Anhui, China

R. Jia (✉)
Chongqing Institute of Green and Intelligent Technology, Chinese Academy of Sciences, Chongqing, China
e-mail: jiaruoran@foxmail.com



Y. Ma (ed.), *Proceedings of 4th International Conference on Intelligent Systems Design and Engineering Applications*, Lecture Notes in Electrical Engineering 1478,
https://doi.org/10.1007/978-981-95-1802-9_5

## 1 Introduction

Etching is a critical process in semiconductor manufacturing, which is a crucial step in the translation of design layouts into wafer results, as its quality has a direct bearing on the performance of chips or integrated circuits. With an accurate etch bias model, the etch process can be meticulously evaluated, simulated, and optimized, thereby enhancing the quality of the etched features and mitigating manufacturing costs. Despite the widespread application of plasma etch processes in semi-conductor manufacturing, modeling the formation of patterns remains a challenging task.

The etch process holds paramount significance in ensuring precise critical dimension (CD) control, particularly for nodes below the 20nm (nanometer) threshold. The etch bias, typically ranging from 3 to 10 nm, is a critical factor for 20nm node devices, often comparable to the proximity effect introduced by the lithographic process. Given the intricate nature of the etch process, an empirical etch bias model has historically been employed, particularly in the context of Optical Proximity Correction (OPC). In contrast, the lithographic model enjoys a closer alignment with the 1st principal model, which is firmly grounded in physics and chemical reactions.

Reactive ion etching (RIE) is particularly susceptible to a wide range of factors, including the nature of the plasma, the substrate material, chamber configuration, gas composition, gas flow rate, and more. However, precise modeling of the etch behavior with respect to plasma and chamber structure design, as well as the properties of different film materials, cannot be accurately simulated in the short term.

Currently, the available etch bias models are primarily focused on analyzing the pattern geometry variations of different layout characteristics under fixed etch recipes and with a fixed film structure and chamber. In those compact etch bias models, modeling is primarily centered on layout characteristics information such as pattern density and spacing between patterns, while other factors are deemed as constant and in-variable. Nevertheless, this modeling approach gives rise to limited generalization capabilities. Consequently, whenever there is a variation in process parameters, such as modifications in film information or etch recipes, the model necessitates the collection of new online data and a subsequent remodeling process.

Among the layout characteristics information, the etch aperture effect and loading effect, which significantly influence the etch quality (known as etch bias), are de-fined as the discrepancy between the post-lithography contour and the post-etch contour. Accurate consideration of these variables and effects is essential for precise etching and optimal semiconductor manufacturing outcomes [1–6].

As observed in Fig. 1, etch bias is essentially the difference between an ADI (After Develop Inspection) polygon and an AEI (After Etch Inspection) polygon. From a layout perspective, we can define etch bias as a directed line segment, with its starting point located on the ADI polygon and its endpoint on the AEI polygon. The direction of etch bias is fixed as the perpendicular direction to the curvature of the starting point, which we define as the etch direction, as shown in Fig. 2. Therefore, the task of the etch bias model can be abstracted as a machine learning regression problem, namely predicting the distance value along the etch direction from each point on the ADI polygon to the AEI polygon.

The Variable Etch Bias (VEB) model has emerged as a widely used etch bias model for OPC. It primarily considers pattern geometry as a crucial factor influencing loading and aperture effects in its approximations to predict etch bias. For a given etch-process condition, VEB evaluates etch bias through the convolution of pattern geometry with multiple fixed-radius kernel functions. This approach restricts the local plasma distribution based on the specific geometry, resulting in more precise predictions.

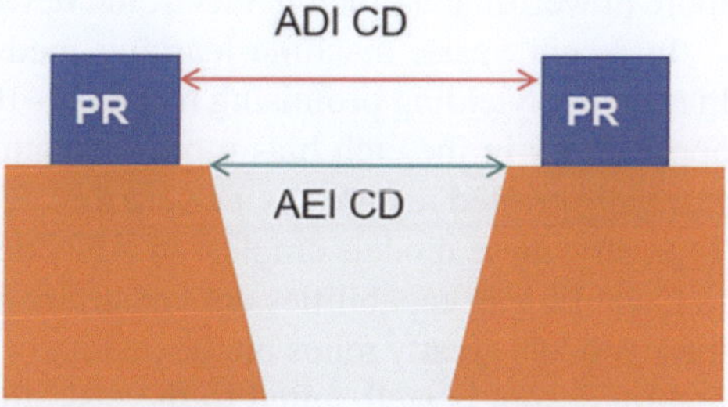


**Fig. 1** Image of the 1D cross-section of etch bias

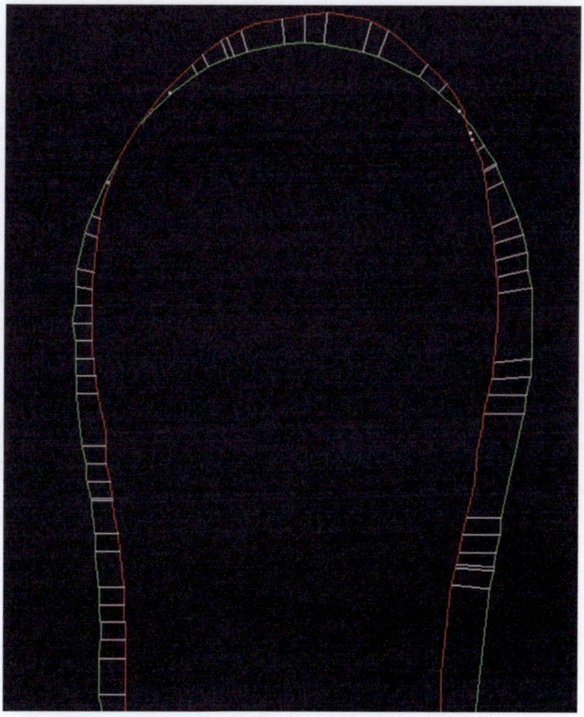

**Fig. 2** An example Image of etch bias in the layout

Although the Variable Etch Bias (VEB) model has achieved remarkable success in its applications, etch bias model still requires further enhancement as feature dimensions continue to shrink below the 20nm node. The complexities arising from increasingly intricate patterns and tighter tolerances necessitate a more sophisticated etch bias model capable of capturing the nuances of the etch process with enhanced precision.

As technology advances, the etch process remains a crucial aspect of semiconductor manufacturing. The pursuit of improved etch models is not merely about enhancing the quality of etched features; it also involves optimizing the overall manufacturing process, reducing costs, and ultimately enabling the production of more powerful and efficient electronic devices.

In recent years, machine learning methods have been integrated into the etch bias model, yielding promising results [7–10]. Nevertheless, these machine learning applications in the etch bias model remain semi-empirical in nature. They rely on manually crafted features, which are then fed into artificial neural networks (ANNs). Typically, these models employ an ANN network structure, yet their feature extraction and fit-ting capabilities are constrained. As a result, the overall performance of these models greatly relies on the quality of the manually crafted features. When the feature design is well-suited to the task, the model performs optimally; otherwise, its performance can be rather substandard.

With the continuous development of AI technology, more AI models with stronger feature extraction and fitting capabilities have been proposed. In the field of etch bias modeling, there have also been attempts to use more complex models to extract key features and fit etch bias using data, rather than relying solely on manually designed kernel functions [10]. The ASML-Brion etch model employs a modeling flow with image input for a deep-learning Convolutional Neural Network (CNN). However, this method also neglects the consideration of key influencing factors in the etch process, such as aperture and load effects.

In this paper, we propose a bilateral-branch adaptive micro-load radius sensing etching bias prediction model. This model automatically senses the best distance of the micro-load effect based on pattern characteristics and combines it with the macro-load effect to predict the etching bias of the current pattern with greater accuracy.

In summary, the key contributions of this paper are as follows:

Firstly, we demonstrate that the micro-load effect significantly impacts the performance of the CNN model when the input size varies, which highlights the importance of capturing fine-grained etching characteristics.

Secondly, due to differences in pattern density, the optimal Field of Vision (FOV) for capturing the pattern micro-load effect differs, emphasizing the need for an adaptive approach.

Thirdly, we introduce the Bilateral-Branch Network Adaptive FOV Etch Model, which comprises two distinct branches. The micro-load branch automatically determines the optimal distance for extracting micro-load characteristics based on pattern features, while the macro-load branch analyzes density characteristics across a broad range of images. By combining the characteristics of both macro and micro load effects, our model can predict etching bias more precisely.

The final experimental analysis reveals that our model outperforms existing CNN-based models in terms of accuracy and robustness.

## 2 Related Work

### 2.1 Traditional Empirical Etch Model

Rule-based etch model, adjusting contour via a look-up table of etch bias, has been adapted in a few technology nodes, but simple rules alone cannot accurately correct various pattern shapes.

In the VEB model [3, 8], a famous model-based etch model, etch bias is modeled as a function of a few handmade empirical parameters:

$$Bias = A_0 + \sum_{i=1}^{N_{\text{Den}}} B_i \text{Den}^i + \sum_{j=1}^{N_{\text{Vis}}} C_j \text{Vis}^j + \sum_{k=1}^{N_{\text{Blo}}} D_k \text{Blo}^k \tag{1}$$

where Den, Vis, and Blo are the respond value of a density kernel, a visible kernel, and a blocked kernel, respectively; $A_0$, $B_i s$, $C_j s$, $D_k s$ are fitting coefficients determined through linear regression.

Figure 3 illustrates a widely adopted sampling method within the realm of space tracing and numerous other applications, which samples a grid through multiple circles sharing the same center [11]. Such a method proves effective in dissecting the geometric variances of patterns possessing diverse characteristics, a factor of paramount significance in establishing the etch bias model.

Although rule-based and model-based etch models have been successfully applied in earlier technology nodes, they become insufficient as the technology node shrinks [8]. It is expected that, in 10-nm technology, on-chip variation (OCV) is still 28% beyond the tolerance even though rule-based etch model is applied [12], and OCV is still 15% of gate size by using model-based etch model in 20nm DRAM devices [13].

### 2.2 Machine-Learning (ML) Etch Model

Machine learning approaches have been introduced to optical proximity correction (OPC), and the etch model is particularly intriguing from these perspectives [8]. Firstly, ML is an empirical model that does not fully comprehend the underlying physics of a given ML domain. Additionally, through experience, we have a general under-standing of which process parameters are crucial for etch bias. Therefore,

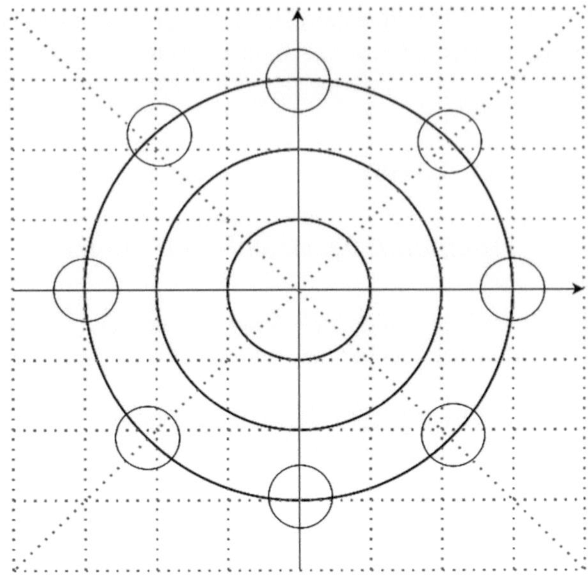

**Fig. 3** Concentric circle sample method

developing an etch ML model is a challenging yet rewarding task for accurately predicting etch bias.

However, the primary challenge of machine learning models lies in the fact that the key determinant of a model's generalization ability is not the model itself but rather its input. If a model is fed with as many relevant parameters as possible, it becomes easier for the model to produce better results. Nevertheless, determining how to calculate and select these input parameters heavily relies on personal experience.

Rui Chen et al. built a etch bias prediction model based on machine learning, they extract features from the layout to build the training dataset and divide each sample into N clusters according to similarities by principle component analysis [7]. They train the ANN model for each individual cluster, which achieves absolute error less than 4 nm with a standard deviation of 3nm and average relative error of 15% for 2D pattern.

Youngchang Kim et al. use simulated CD from conventional VEB model as an input of Neural Network, they selected 4 groups of parameters in this work and 5 different combinations of the inputs are investigated to figure out which are dominant for ML modeling [10]. Finally, the accuracy and predictability of the ML model are improved from 2.4/2.6 to 0.65/1.34 nm in test case of 10 nm metal.

In conclusion, the ultimate goal of the etch bias model is to eliminate its reliance on manually designed features. Instead, the model should autonomously learn features from data and establish a precise input–output relationship. This approach lays a solid foundation for developing a highly accurate etch bias model.

In general, there is no difference between the etch bias model of machine learning and the VEB method in acquiring layout features. Both adopt some artificially experience-related kernel functions, but the basic linear regression model is replaced with a more powerful ANN or other models.

### 2.3 CNN-Based Etch Model

With the advancement and widespread application of AI (Artificial Intelligence) technologies in the semiconductor industry, increasingly complex and robust models have been incorporated into etch modeling. ASML-Brion introduced a CNN-based framework specifically designed to predict etch bias [9].

They use massive reliable AEI SEM images obtained from traditional CD-SEM flow and the extracted CD gauges and EP gauges to describe all possible etch process effects and train the deep-learning Etch model. Leveraging its exceptional feature extraction capabilities, the CNN neural network automatically learns the key factors that influence the etch bias of surrounding graphics within the image, ultimately leading to accurate etch bias predictions.

The significant advantage of this method lies in its ability to eliminate the reliance on expert-derived feature functions. Instead, etch bias is calculated through the feature extraction and fitting capabilities of the sophisticated deep neural network. The input image incorporates the etch bias point that needs to be predicted.

However, there are certain limitations to this approach. Some factors that are crucial in the actual manufacturing process are overlooked. For instance, there's a need to determine the optimal size of the input image for the CNN model to accurately capture the micro-loading effect. Additionally, the framework must consider whether different ranges of loading effects should be considered for varying patterns. Addressing these challenges is essential for further enhancing the accuracy and reliability of the etch bias prediction model.

## 3 FOV-Adaptive Bilateral-Branch Network Etch Model

### 3.1 Loading Effect

The loading effect is a phenomenon in which the etch rate varies depending on the size of the etchable surface area exposed to the etchant. This effect is particularly prominent in chemical reactions and operations throughout the etching process. Macro and micro loading effects are the primary types of loading effects observed during etching, both having a significant impact on the etch rate and etch bias. The etch rate is directly proportional to the depth of the etched trenches, while the etch bias represents the difference between the critical dimension (CD) of the etched pattern and the CD of the photoresist (PR) pattern.

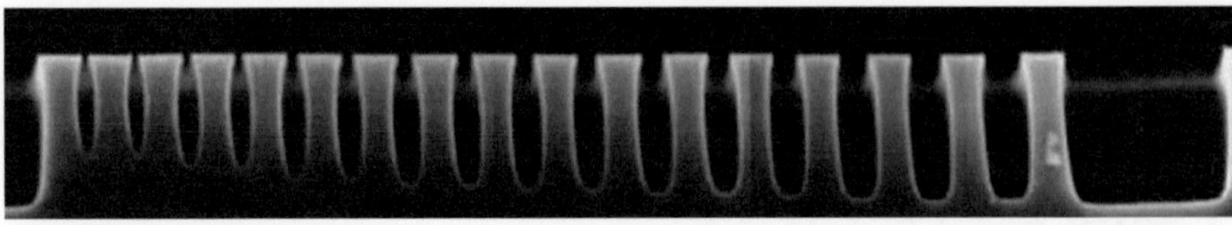

**Fig. 4** The loading effect impacts the etch rate, resulting in variations in etch depth and CD

More specifically, the macro-loading effect leads to a decrease in the etch rate in regions with a larger etchable surface area. Conversely, the micro-loading effect causes a decrease in the etch rate in areas where a higher number of trenches require etching. Additionally, both macro and micro loading effects contribute to variations in the etch bias, potentially affecting the dimensional accuracy and uniformity of the etched features. This is because the probability of etchant particle reflection is higher in regions with narrower and more densely packed etched trenches.

Based on the physical characteristics of the etching model, the etching effect is mainly affected by the aperture and the loading effect. The aperture effect is mainly related to the film stack and the etch steps. The loading effect refers to the decrease in etching rate or uneven distribution caused by the consumption of local etching gas exceeding its supply. The loading effect can be classified into three types: macro-loading, micro-loading, and aspect ratio dependent etching, which is related to the etching depth-to-width ratio. Among them, macro-loading and micro-loading have a significant impact on the critical dimension (CD) of the substrate in the layout [14].

From a layout perspective, the loading effect refers to the distribution of patterns within a local area. As illustrated in Figs. 4 and 5, different pattern densities result in variations in etch rates and etch bias. For etch bias models based on layout, the primary objective is to identify a modeling approach that effectively captures the loading effect. Research an efficient and accurate method to calculate all key influencing factors for the etch bias of each point on the ADI contour, such as the shape of the pattern itself, the position of this point within the pattern, the shapes and densities of surrounding patterns, and the effective influence range of surrounding patterns on this point.

## *3.2 The Effect of Input Size on CNN Model*

When utilizing CNN networks to construct an etch bias model, it is crucial to consider the dimensions of the input images and the corresponding observation ranges within the patterns (When converting the layout into an image, we assume that 1nm corresponds to one pixel by default, image size. Therefore, the size of the image represents the range of surrounding patterns considered in modeling). This is because the receptive field size of a CNN network is typically fixed during its design phase. Typically, meticulously selecting the ratio of image size to observation range

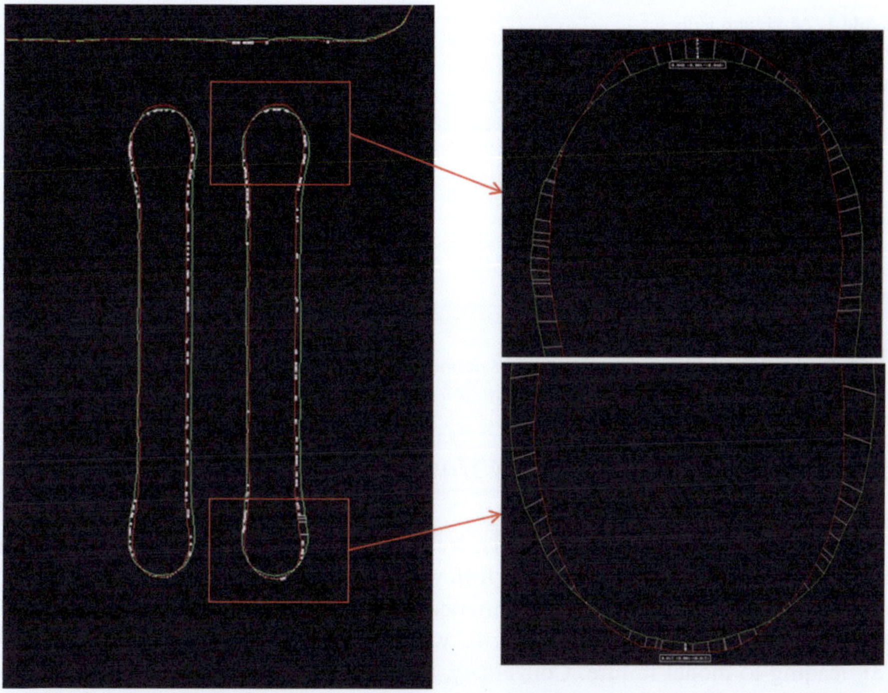

**Fig. 5** Different loading effects have distinct impacts on etch bias

can mitigate distortions arising from resizing input images and enhance the model's performance. Additionally, the input size of images is constrained by the computing and storage capabilities of the devices where the CNN-based model is deployed.

To investigate the influence of various observation ranges on the prediction accuracy of the etch bias, using the same network and dataset, the following experiments were conducted. Using 56nm as the unit, the observation FOV was gradually expanded, and subsequently, the accuracy of the CNN-based etch bias model was analyzed under different FOV radii. In this study, Mean Absolute Error (MAE) and Contour Edge Placement Error (EPE) were employed as evaluation metrics to assess the model's performance as Table 1 shows.

Table 1 demonstrates that the FOV radius has a notable impact on the accuracy of the etch bias prediction using CNNs. When a smaller FOV radius is utilized, such as 56nm or 112nm, the MAEs of the CNN-based model exceed 2nm. However, as the FOV radius increases, the MAEs of the CNN-based model decrease significantly to approximately 1.6nm and subsequently fluctuate within the range of 1.6–1.8 nm. Furthermore, the average EPE from real AEI contours to predicted AEI contours exhibits a similar trend as the average MAE from ADI contours to predicted AEI contours. This implies that the MAE is a valid and effective metric for evaluating etch bias models.

**Table 1** MAE and EPE under different FOV radii

| FOV radius (nm) | Avg. MAE (nm) | Avg. EPE (nm) |
|---|---|---|
| 56 | 2.85 | 2.79 |
| 112 | 2.10 | 2.17 |
| 168 | 1.66 | 1.65 |
| 224 | **1.66** | **1.60** |
| 280 | 1.75 | 1.77 |
| 336 | 1.72 | 1.68 |
| 392 | 1.84 | 1.83 |
| 448 | 1.78 | 1.72 |

Bold values define the best fov of different density, which convient for reader to find out the best condition

## *3.3 Different Pattern Density Has Different Optimal FOV*

Based on the aforementioned analysis, we discovered that the trend of the average MAE, specifically its initial decrease followed by fluctuation within a narrow range, actually mirrors the influence of pattern density on model performance. As the FOV radius decreases, the density of patterns within the FOV gradually approaches 0.5 [3], leading to model failure. Conversely, as the FOV radius increases, the density varies accordingly. Furthermore, different patterns exhibit varying interaction ranges and strengths with surrounding patterns. Some impacts are short-range, while others ex-tend to longer ranges, which can be captured by adjusting the FOV radius.

Although some current algorithms, both VEB and machine learning-based, have accounted for this effect, the radius of FOV still relies on manual adjustment. For instance, in the VEB model, the kernel radius necessitates manual tweaking. This manual setting not only complicates the operational process but also introduces potential in-stability due to human factors. Consequently, the challenge remains how to automatically and precisely determining the radius of FOV, which is a crucial aspect to address in current algorithm research. In this paper, our objective is to devise a deep learning-based etch bias model that can adaptively ascertain the impact range of micro-loading by automatically seeking an optimal FOV radius. We intend to delve into the correlation between pattern density and the optimal micro-loading FOV radius, and further investigate whether there are additional factors that influence the optimal FOV radius, such as whether 1D graphics and 2D patterns exhibit distinct optimal FOVs under the same pattern density.

To determine the optimal FOV radii across varying pattern densities, we conducted a series of rigorous experiments. Initially, we uniformly partitioned the entire density range of [0.0, 1.0] into 10 distinct intervals and subsequently categorized patterns into 10 corresponding classes based on these density intervals. Subsequently, we evaluated the average MAEs within each density interval for different FOV radii and tabulated the results in Table 2.

**Table 2** MAE distribution on different pattern densities and FOV radii

| Density | FOV | | | | | | | |
|---|---|---|---|---|---|---|---|---|
| | 56 | 112 | 168 | 224 | 280 | 336 | 392 | 448 |
| [0.0–0.1) | N/A | 6.28 | 8.07 | 7.57 | **5.66** | N/A | N/A | N/A |
| [0.1–0.2) | 4.06 | 2.77 | **2.62** | 2.67 | 2.70 | 2.91 | 2.94 | 3.18 |
| [0.2–0.3) | 2.81 | 2.26 | 1.91 | 2.03 | 1.85 | **1.84** | 2.14 | 1.98 |
| [0.3–0.4) | 2.42 | 1.90 | 1.68 | **1.65** | 1.75 | 1.76 | 1.80 | 1.76 |
| [0.4–0.5) | 2.98 | 2.11 | **1.46** | 1.52 | 1.83 | 1.53 | 1.76 | 1.61 |
| [0.6–0.7) | 4.59 | 2.76 | 1.41 | **1.28** | 1.35 | 1.35 | 1.48 | 1.40 |
| [0.7–0.8) | 3.44 | N/A | 1.69 | **0.72** | 0.97 | 0.73 | N/A | N/A |
| [0.8–0.9) | N/A | N/A | N/A | N/A | N/A | N/A | N/A | N/A |
| [0.9–1.0] | N/A | N/A | N/A | N/A | N/A | N/A | N/A | N/A |
| [0.6–0.7) | 4.59 | 2.76 | 1.41 | **1.28** | 1.35 | 1.35 | 1.48 | 1.40 |

Bold values define the best fov of different density, which convient for reader to find out the best condition

Table 2 demonstrates that the optimal FOV radii vary significantly based on different pattern densities ("N/A" indicates that there are no samples available for analysis under the current conditions). Notably, across all pattern densities, the optimal FOV radii are no smaller than 168nm. Furthermore, in the majority of cases (exceeding 60%), the optimal FOV radius is either 168 nm or 224 nm, aligning with the findings presented in Table 1. Consequently, as a general rule, setting the FOV radius within the range of 168–224 nm can ensure that the model achieves satisfactory performance across a wide variety of patterns with differing densities.

## *3.4 Etch Macro-Loading Effect*

The macro-loading effect is another type of process variation that arises due to the disparities in pattern densities between isolated and dense pattern areas [15].

During our experiments, we observed an unexpected phenomenon: for certain patterns, despite the ADI contours being relatively close, the AEI contours exhibited significant differences, as depicted in Fig. 6, with discrepancies reaching up to 4nm. The short-range micro-loading effect alone appears insufficient to explain this occurrence. Consequently, in this paper, we argue that the etch bias model ought to incorporate the long-distance macro-loading effect to enhance the model's precision in capturing finer details.

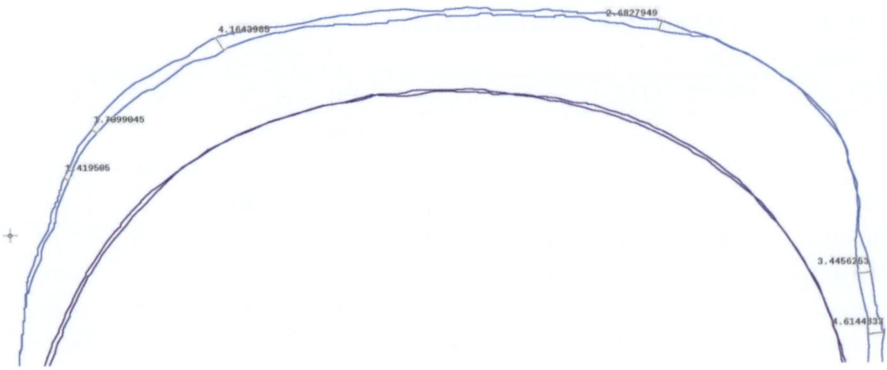


**Fig. 6** Stacked ADI (purple) and AEI (navy) contours

### *3.5 FOV-Adaptive Bilateral-Branch Network*

This paper introduces a novel FOV-Adaptive Bilateral-Branch Network (FA-BBN) designed to model etch bias, specifically the discrepancy between ADI and AEI contours. Its captivating architecture, drawing inspiration from [16], features bilateral branches that separately accommodate short-range micro-loading images and long-range macro-loading images. Within the micro-loading branch, an attention block plays a pivotal role in predicting the optimal variance for a Gaussian Kernel, based on the pattern's feature maps. Subsequently, a Gaussian mask, crafted from this optimal variance, akin to setting an ideal field of view (FOV), is multiplied with the extracted feature maps, yielding Gaussian-weighted feature maps. This Gaussian mask significantly impacts our model's performance, as its decaying weight, extending outwards from the center, aptly represents the etch behavior of radicals. Conversely, the macro-loading branch solely comprises a CNN for feature extraction. Once the bilateral branches have processed their inputs, the flattened feature vectors are merged and fed into a fully-connected network, ultimately outputting the predicted etch biases.

The overall FA-BBN architecture is depicted in Fig. 7. It contains three main components:

- the micro-loading branch to extract a compact micro- loading feature representation;
- the macro-loading branch to extract macro-loading feature representation;
- the feed-forward full-connected network that makes the final etch bias prediction.

The micro-loading branch inputs a picture of 224 * 224 pixels with a small-scale observation range of 224 * 224 nm, where 1 pixel represents 1 nano meter. Figure 8 depicts the architecture of the Gaussian attention in the micro-loading branch. First, it extracts features through a CNN backbone and divides it into two branches. One branch predicts the Gaussian kernel radius and Gaussian model variance through

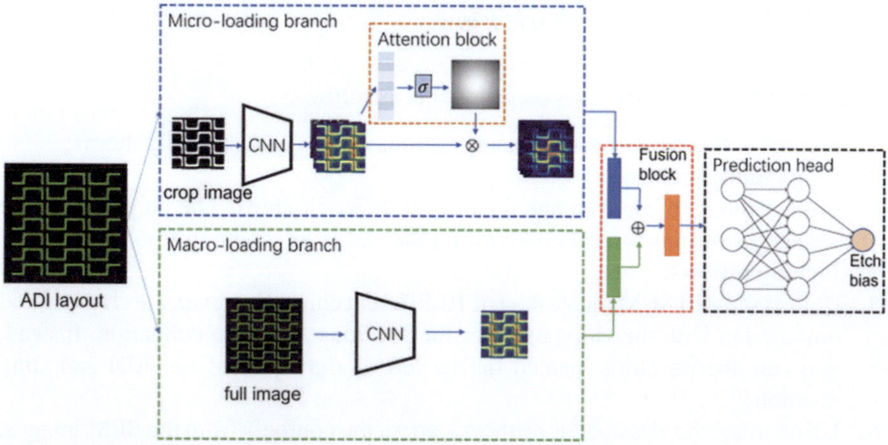


**Fig. 7** The architecture of FA-BBN

the fully connected layer, and generates an attention map based on the above two parameters. After combining the attention map with the original feature map, the FOV adaptive feature map is obtained, and then the feature vector of the micro-load branch is output through the fully connected layer.

The macro-loading branch inputs an image of 224 * 224 pixels with a large-scale observation range of 896 * 896 nm, and outputs a macro-loading feature vector after passing through a CNN backbone and a fully connected layer. This branch is used to model the impact of pattern density on the final result of the etch bias model.

In the forward prediction part, after fusing the input micro-load and macro-load feature vectors, the final etch bias prediction value is obtained after passing through several fully connected layers.

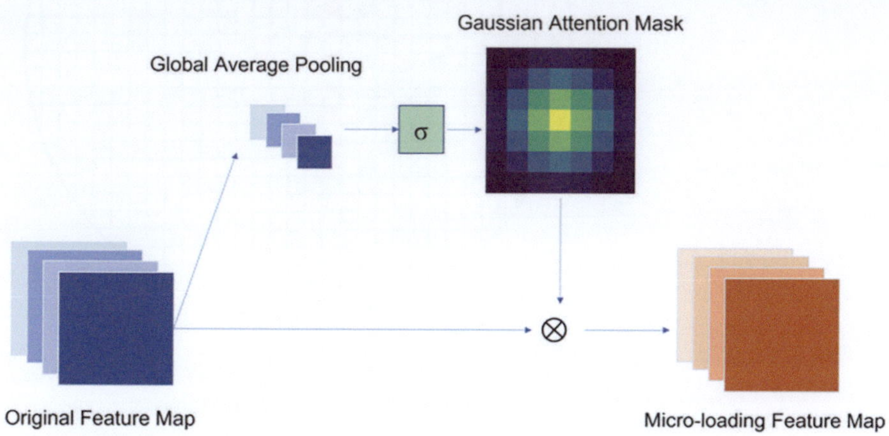


**Fig. 8** The architecture of the Gaussian attention in the micro-loading branch

## 3.6 FA-BBN Network Work Flow

The entire workflow of the etch bias model is as follows:

1. Prepare a Test Mask encompassing various shapes and sizes of patterns;
2. Lithography process;
3. Collect the ADI SEM image data of 10 different chips, The chips should be located in various typical positions on the wafer, as shown in Fig. 9;
4. Etching process;
5. Collect the AEI SEM image data of 10 different chips, However, the chips should not overlap with the chips used for the previous ADI data collection. Instead, you can choose chips located on the left or right side of the ADI sampling positions;
6. Using image processing algorithms, extract the contours from the SEM images;
7. Combine the contours extracted from different chips and perform an averaging operation to obtain a smoothed contour as depicted in Fig. 10. The aim is to eliminate random errors;
8. Using Polygon as the unit, divide all Pattern contours into a training set and a validation set in a 9:1 ratio based on quantity;
9. Train FA-BNN network;
10. Validate FA-BNN network.

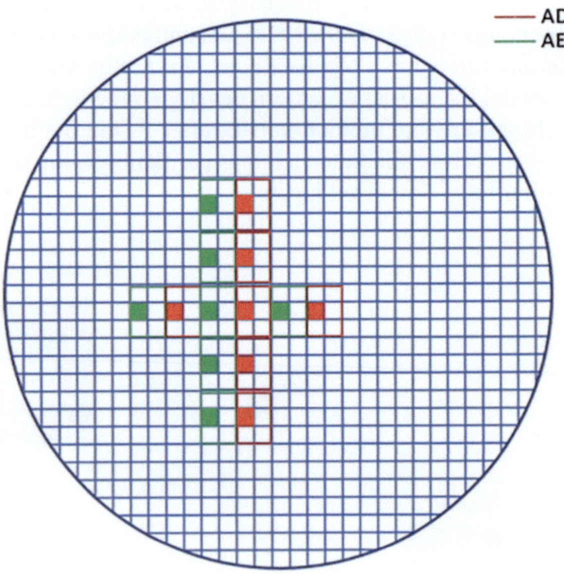


**Fig. 9** The chips selected for the SEM image data collection

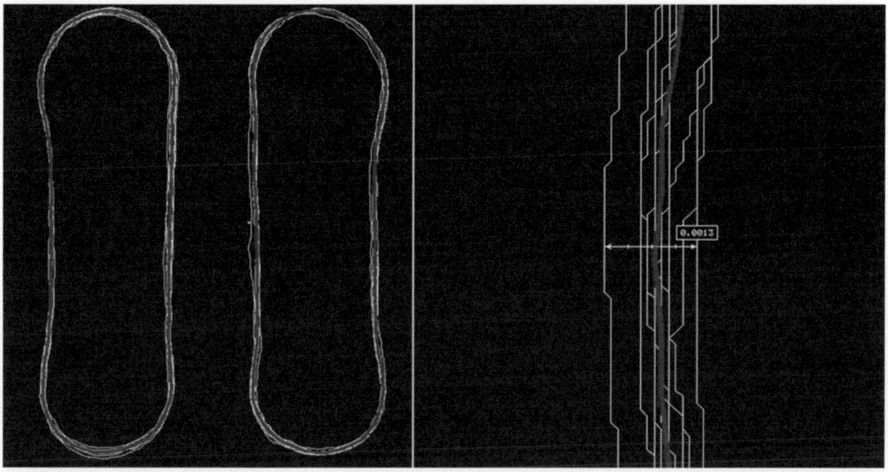


**Fig. 10** The contours from different chips and obtained by performing an averaging operation

# 4 Experiments Data

Traditional synthetic patterns and complex actual patterns are both used in our experiments. The traditional synthetic patterns are composed of parametric patterns, e.g., the dense line and space (DLS), isolated line (Iso-line), and line-end to line-end (E2E), as shown in Fig. 11. The advantage of using synthetic patterns is that they are easy to customize and can depict the trend of etch bias with regard to the change of geometry characters, e.g., critical dimension (CD) and space of DLS.

We also use complex patterns deriving from real test layouts as validation dataset. The test layouts are prepared by shrinking some actual layouts produced via earlier technologies or synthesizing design layouts via a preliminary version of the new technology library.

After being processed by OPC, lithography, and etching process, all patterns are transferred to photoresists and then to wafers. We collected 10 ADIs and the corresponding AEIs from the same sampling area in 10 different dies on a wafer, then extracted all contours and performed an averaging operation to reduce the error caused by noise and disturbance to the model. Overall, we obtained 1882 paired ADI-AEI polygons, further building the training set with 1641 polygons containing 1.6 million points and the testing set with 241 polygons containing 0.2 million points.

Furthermore, we transformed all polygons in GDSII file into binary bitmaps where the inner areas of polygons are labeled by 1, while the outer areas of polygons are labeled by 0. As shown in Fig. 12, each bitmap corresponds to one sampling point that is located in the center of bitmap. Meanwhile, the observation range of each bitmap with the same image resolution enlarges from left to right. This transform is performed for all sampling points and enables the image-based deep learning framework to deal with ADI contours.

**Fig. 11** Synthetic patterns used to train model

**Fig. 12** ADI bitmaps with 224 * 224 pixels under different FOVs, FOV radius is 56 nm, 112 nm, 224 nm, 448 nm, respectively

## *4.1 Technical Details*

The backbone of the FA-BBN is ResNet [17] which has achieved significant success in the computer vision field. ResNets are good at learning the differences in similar images so that we can find out the tiny differences in a large number of repeat patterns. We use ResNet-18 and ResNet-50, respectively, to automatically extract features from bitmaps that are generated from ADI contours. After passing the average pooling layer, the feature vectors of both the micro-loading branch and the macro-loading branch are merged by using a Weighted Feature Fusion (WFF) block. Then, the weighted feature vector is used to predict the etch bias via a full-connected network.

We train the proposed FA-BBN via SGD optimizer with an initial learning rate, 0.0001, and weight decay 0.001. The initial fusion weight of each branch is set to 0.5. From Fig. 13, we can see that the model has good convergence. With the increase in the number of iterations, the loss function of the model will gradually decrease to a relatively stable value.

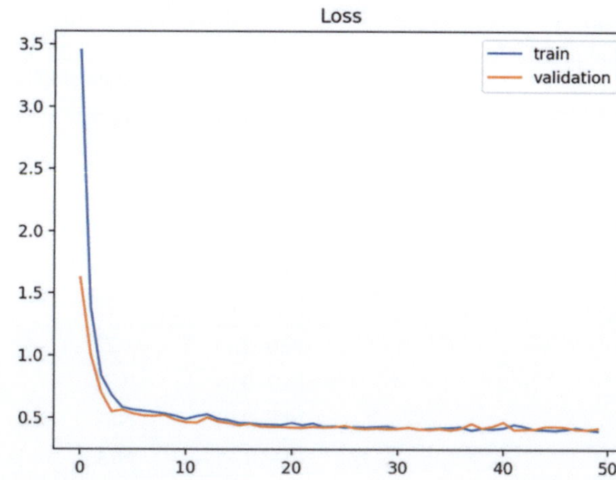


**Fig. 13** Train and validation loss change trend

## *4.2 Analysis*

In order to verify the effectiveness of the FA-BBN etch bias model, this study randomly takes 100 sampling regions on the real chip to verify the model performance. As shown in Fig. 14, by comparing the predicted and measured the etch bias, they do exhibit the similar trend, in which each individual predicted value does not differ too much from the corresponding measured value, excluding few exceptions showing relatively larger gap.

In addition, we compared the absolute error of predicted etch biases and the corresponding measured etch biases in Fig. 14. Except for a few points, the absolute errors of most points stay within 5 nm. The calculated mean value of these errors is around 1.9 nm with a standard deviation of 1.85 nm.

Table 3 shows that the proposed FA-BBN outperforms the model by using the manual FOV adaption. The average MAEs of predicted etch bias of FA-BBN are in

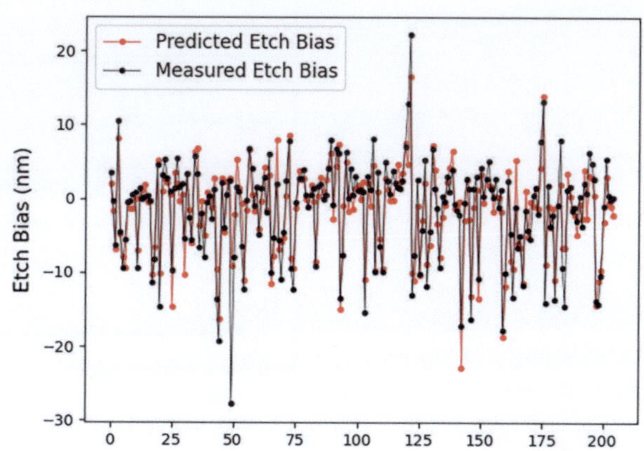


**Fig. 14** Comparison of predicted and measured etch bias

**Table 3** Average MAEs Comparison Between FA-BBN and Manual FOV Adaption

| Density | Method | |
|---|---|---|
| | Manual FOV adaption | FA-BBN |
| [0.0–0.1) | 5.66 | 2.59 |
| [0.1–0.2) | 2.62 | 2.60 |
| [0.2–0.3) | 1.84 | 1.52 |
| [0.3–0.4) | 1.65 | 1.49 |
| [0.4–0.5) | 1.46 | 1.39 |
| [0.5–0.6) | 1.15 | 1.08 |
| [0.6–0.7) | 1.28 | 1.18 |
| [0.7–0.8) | 0.72 | 0.71 |
| [0.8–0.9) | N/A | N/A |
| [0.9–1.0] | N/A | N/A |

the range from 0.71 to 2.60 nm, and more stable than those of models adapting FOV manually.

By using the proposed FA-BBN etch model, we predicted AEI contours according to the ADI contours. The predicted AEI contours are shown in Fig. 15 and the corresponding the histogram of EPE between the predicted contours and the real contours are shown in Fig. 16. For our FA-BBN etch model, about 70% of EPE values are below 2 nm and about 85% of EPE values are below 3 nm.

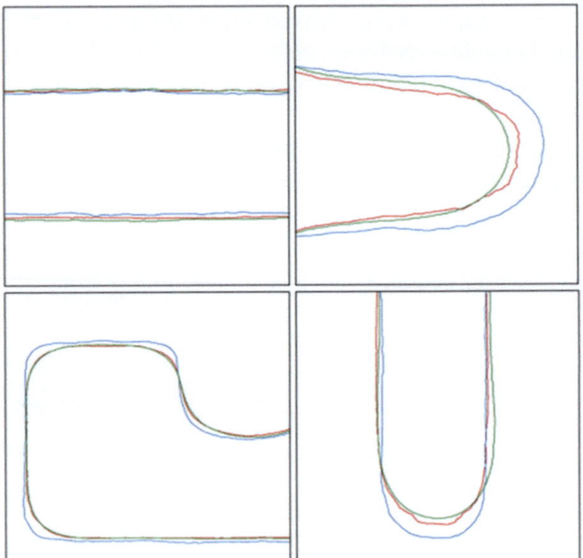

**Fig. 15** Relative position of predicted AEI contours (red) to ADI contours (blue) and AEI contours (green)

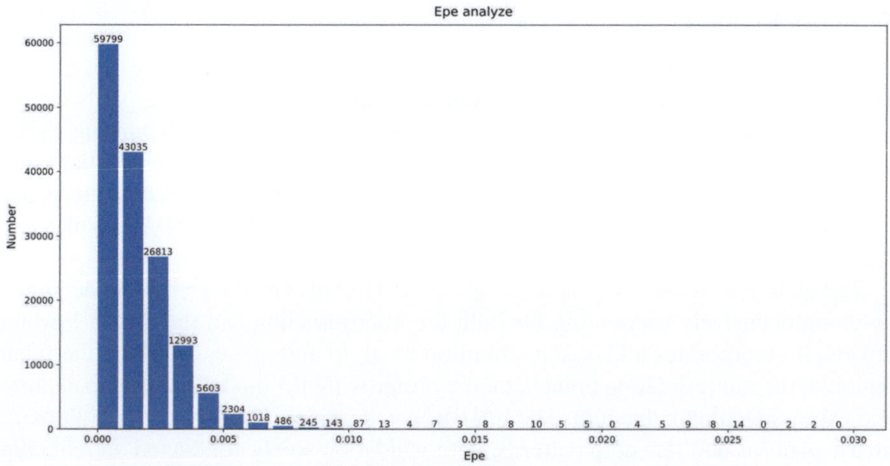


**Fig. 16** Histogram of EPE between the predicted contours and real contours

In addition, an ablation study was performed to demonstrate the validation of the Gaussian attention block and the bilateral-branch architecture, we train 3 networks as follows:

- a basic network that only uses a modified resnet-18;
- an attention-based network that adds a Gaussian attention block to the basic network;
- a FA-BBN with a micro-loading branch and a macro-loading branch.

The averaged mean absolute error (MAE), $R^2$ and averaged edge placement error (EPE) are calculated, respectively, and the results are shown in Table 4. The experimental results indicate that the proposed FA-BBN network outperformed both the basic network and the attention-based network in MAE and EPE. Yet, it exhibits only a marginal advantage in $R^2$. Furthermore, we observed that the attention-based network also exhibits improvements in all performance metrics compared to the basic network, suggesting that the Gaussian attention block significantly contributes to the enhancement of the model's performance.

**Table 4** Performance comparison among different models

| | ResNet-18 | ResNet-18 + Gaussian attention | FA-BBN |
|---|---|---|---|
| MAE (nm) | 1.87 | 1.72 | 1.64 |
| $R^2$ | 0.85 | 0.87 | 0.88 |
| EPE (nm) | 1.82 | 1.67 | 1.43 |

## 5 Conclusions

This paper systematically analyzes the impact of different FOVs on the performance of the etch bias model for the first time. By modeling patterns with varying FOVs based on density attributes, we calculate the model performance and reveal that each pattern type has an optimal FOV. Based on the above experimental results, this paper proposes a novel FOV-adaptive etch bias model, termed the FA-BBN (Field-of-View Adaptive Bilateral-Branch Network) model.

The etch bias model we proposed significantly enhances the prediction accuracy by comprehensively accounting for both the micro-loading and the macro-loading effects. It incorporates a Gaussian attention block to automatically adjust the focal region of the micro-loading branch, thereby improving the model's ability to capture and express etching behaviors. The FA-BBN etch model is trained using both parametric patterns and real chip patterns, with validation solely conducted on real chip patterns to facilitate the deployment of our model into real-world applications. Experimental results confirm the effectiveness of the Gaussian attention mechanism in improving etch bias prediction accuracy and robustness.

Despite its advancements, our method currently faces the issue of excessive computational cost, as the model requires generating an image for each point on the contour during the neural network training process. Additionally, the model only models the numerical value of etch bias, with the etch direction assumed as the perpendicular line to the curvature, which often does not align with the actual deviation direction. Future research should aim to develop a unified model capable of predicting both the movement direction and magnitude of the ADI contour, which would significantly contribute to enhancing the etch bias modeling methodology.

## References

1. Proximity compensation In: Proceeding of SPIE **6520**, 652046 (2007) [Online]. Available: https://doi.org/10.1117/12.711841
2. Granik, Y., Toublan, O., Cobb, N., Sahouria, E., Donnelly, T., Do, T.: Chip-scale simulation of pattern processing: methods, models, accuracy, and applications. In: 17th VMIC Conference Proceedings, pp. 249–253 (2000)
3. Granik, Y.: Correction for etch proximity: new models and applications. In: SPIE Advanced Lithography (2001)
4. Granik, Y.: Dry etch proximity modeling in mask fabrication. In: Proceeding of SPIE, vol. 5130, 08/01 2003. https://doi.org/10.1117/12.504052
5. Sunwook, J., Thuy, D., John, S.: Finding practical phenomenological models that include both photoresist behavior and etch process effects. In: Proceeding of SPIE, vol. 9428, p. 94280I (2015) [Online]. Available: https://doi.org/10.1117/12.2087096
6. Cobb, N.B., Granik, Y., Sahouria, E.Y., Toublan, O.: Full-chip process simulation for silicon DRC (2000)
7. Chen, R., et al.: ETCH model based on machine learning. In: 2020 China Semiconductor Technology International Conference (CSTIC), 26 June-17 July 2020 2020, pp. 1–4. https://doi.org/10.1109/CSTIC49141.2020.9282462

8. Shim, S., Shin, Y.: Etch proximity correction through machine-learning-driven etch bias model. In: SPIE Advanced Lithography (2016)
9. Weina, S., et al.: OPC accuracy improvement through deep-learning based etch model. In: Proc. SPIE, vol. 12056, p. 1205607 (2022) [Online]. Available: https://doi.org/10.1117/12.2613926
10. Youngchang, K., Sunwook, J., DooHwan, K., Vlad, L., Germain, F.: Predictable etch model using machine learning. In: Proc. SPIE, vol. 10961, p. 1096106, (2019) [Online]. Available: https://doi.org/10.1117/12.2515271
11. Myridis, N.E., Chamzas, C.: Sampling on concentric circles [MRI image reconstruction application]. IEEE Trans. Med. Imaging **17**(2), 294–299 (1998). https://doi.org/10.1109/42.700741
12. Mohamed, S., Ayman, H.: Efficient etch bias compensation techniques for accurate on-wafer patterning. In: Proceeding of SPIE, vol. 9427, p. 94270X (2015) [Online]. Available: https://doi.org/10.1117/12.2085956
13. Park, J.-G., Kim, S.-w., Shim, S.-H., Suh, S., Oh, H.-k.: The effective etch process proximity correction methodology for improving on chip CD variation in 20 nm node DRAM gate. Adv. Lithography (2011)
14. Jani, K., Jyrki, K., Sami, F.: Loading effects in deep silicon etching. In: Proc. SPIE, vol. 4174, pp. 90–97 (2000) [Online]. Available: https://doi.org/10.1117/12.396475
15. Yang, Y.M., et al.: Silicon trench etch uniformity improvement for microloading and macro-to-macro loading for sub-14nmNode. In: 2018 IEEE International Interconnect Technology Conference (IITC), 4–7 June 2018, pp. 79–81 (2018). https://doi.org/10.1109/IITC.2018.8430443
16. Zhou, B., Cui, Q., Wei, X.-S., Chen, Z.-M.: BBN: bilateral-branch network with cumulative learning for long-tailed visual recognition. https://doi.org/10.48550/arXiv.1912.02413
17. Li, Z., Liu, F., Yang, W., Peng, S., Zhou, J.: A survey of convolutional neural networks: analysis, applications, and prospects. IEEE Trans. Neural Netw. Learn. Syst. **33**(12), 6999–7019 (2022). https://doi.org/10.1109/TNNLS.2021.3084827

# IoT-Based Smart Agriculture and Crop Recommendation Using ML

**Simran Jaggi, Sunil K. Singh, Sudhakar Kumar, Anish Sharma, Varsha Arya, Kwok Tai Chui, and Brij B. Gupta**

**Abstract** This research presents an agritech system that integrates IoT and machine learning to revolutionize agriculture. IoT sensors monitor critical factors such as soil moisture, temperature, and crop health, collecting data for analysis through cloud computing and ML algorithms. The system provides crop recommendations based on soil composition, climate, and crop performance, enhancing crop yield, optimizing resource use, and promoting sustainability. Key methodologies include IoT sensor deployment, data preprocessing, and ML model evaluation. Wireless sensors capture environmental parameters, transmitting data via IoT-enabled devices to cloud platforms for analysis. ML algorithms process historical and real-time datasets to recommend optimal crops, ensuring enhanced productivity and resource efficiency. Experimental results indicate that the Random Forest algorithm achieved 99% accuracy, while Naive Bayes recorded 98% accuracy in crop prediction. Comparative analysis shows that the system enhances resource efficiency by 25% and improves crop yield

S. Jaggi · S. K. Singh · S. Kumar (✉) · A. Sharma
CSE Department, Chandigarh College of Engineering and Technology, Chandigarh, India
e-mail: sudhakar@ccet.ac.in

S. Jaggi
e-mail: lco22399@ccet.ac.in

S. K. Singh
e-mail: sksingh@ccet.ac.in

A. Sharma
e-mail: co22308@ccet.ac.in

V. Arya · K. T. Chui
Hong Kong Metropolitan University, Hong Kong, SAR, China
e-mail: varya@hkmu.edu.hk

K. T. Chui
e-mail: jktchui@hkmu.edu.hk

B. B. Gupta
Department of Computer Science and Information Engineering, Asia University, Taichung, Taiwan
e-mail: bbgupta@asia.edu.tw

University of Economics and Human Science, Warsaw, Poland

 
Y. Ma (ed.), *Proceedings of 4th International Conference on Intelligent Systems Design and Engineering Applications*, Lecture Notes in Electrical Engineering 1478,
https://doi.org/10.1007/978-981-95-1802-9_6

by 20% over traditional methods. Challenges like IoT accessibility for small-scale farmers and data privacy are acknowledged, encouraging further research.

**Keywords** Crop recommendation · Sustainable agriculture optimization · Crop recommendation · Smart farming · Internet of Things (IoT) · Machine learning · Smart agriculture optimization · Sensors · Digital agricultural revolution

## 1 Introduction

Maintaining the balance of ecosystems while ensuring sustainable food sources depends on sustainable agriculture [1]. With its emphasis on eco-friendly practices, sustainable agriculture supports farmers and conserves resources. It preserves soil quality, minimizes degradation, and encourages water conservation [2]. Additionally, it plays a crucial role in biodiversity preservation and mitigating greenhouse gas emissions. However, achieving sustainability requires innovative approaches that optimize farming while minimizing environmental impact. Traditional agricultural methods have long supported food production, but their effectiveness and sustainability are often compromised. Modern agriculture is essential as the population grows and environmental challenges intensify. Cutting-edge technologies like machine learning (ML) and the Internet of Things (IoT) offer significant potential to enhance agricultural efficiency and sustainability. As seen in Fig. 1, soil pH values have been tracked over time, showing increasing acidity, emphasizing the need for effective land management.

The IoT [3], characterized by interconnected devices capable of collecting and exchanging data, has immense potential in agriculture. By deploying IoT devices such as sensors, drones [4, 5], and actuators, farmers can monitor crucial variables like soil moisture, temperature, and crop health in real time. This enables proactive decision-making, optimizing resource usage and improving yields. Machine learn-

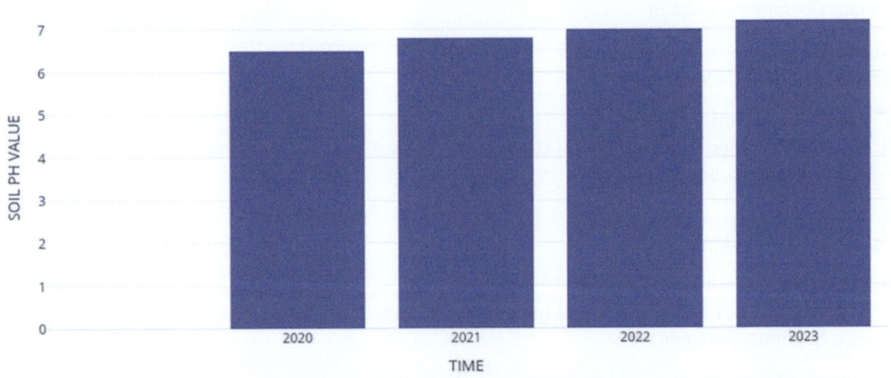


**Fig. 1** Soil health monitoring by soil PH value over time

ing [6, 7], a subset of artificial intelligence [8], complements IoT [9] by analyzing vast datasets to extract insights and make predictions. In agriculture, ML algorithms facilitate crop recommendation and optimization. By analyzing historical data on soil composition, climate, and crop performance, ML model [10] suggest the most suitable crops for specific land parcels, maximizing productivity and sustainability. Blockchain technology in IoT [11] and agriculture enhances security, transparency, and efficiency in various aspects of the agricultural supply chain. The integration of IoT and ML marks a paradigm shift toward smart farming. A real-time system for object detection in agriculture [12] can leverage computer vision and deep learning techniques to enhance efficiency and productivity in farming. These technologies allow data-driven decisions, optimize resource allocation, and enable precision agriculture by applying inputs like water, fertilizer, and pesticides based on real-time analysis. This improves crop yields while reducing environmental impact by minimizing agrochemical use [13]. In countries like India, where agriculture significantly contributes to the economy but faces challenges such as fragmented land ownership and a declining workforce, IoT-based agritech systems offer transformative potential. Initiatives like the "India Digital Ecosystem of Agriculture" leverage technology to digitize farming and enhance resource management. Providing farmers with digital tools and information empowers them to make informed decisions and improve productivity. Figure 2 illustrates various factors related to smart agriculture.

The significance of this research lies in its ability to address critical issues in the agricultural sector, such as the need for sustainable farming practices and the efficient use of resources. By integrating IoT and ML, this study contributes to the field

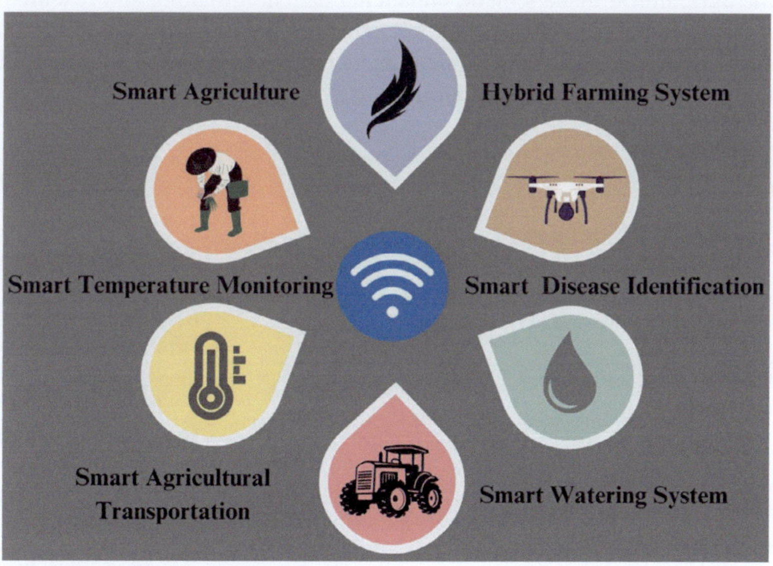


**Fig. 2** The factors related to smart agriculture

by providing innovative solutions that enhance productivity while reducing environmental impact. The originality of this approach is underscored by the novel use of real-time data and advanced algorithms to create a more resilient and efficient farming ecosystem, which is essential for meeting the challenges of a growing population and climate change. In Sect. 2 of this paper, it highlights the details about the previous researchers done in the domain of smart agriculture, Sect. 3 provides a description of the challenges in the existing systems, Sect. 4 explains the detailed methodology of our proposed model, Sect. 5 specifies the results and evaluation of our study and finally, Sect. 6 provides details of conclusion and future scope.

## 2 Literature Review

A thorough examination of the current research endeavors within the domain is imperative to formulate efficient machine learning and deep learning models. In the field of agriculture, the utilization of ML and DL techniques, particularly for image classification, has witnessed a significant rise in popularity in recent times. Consequently, numerous studies have been conducted, employing diverse ML and DL techniques, to construct models for smart farming by leveraging image data. This section of the paper provides a concise summary of several earlier studies on waste segregation, which have employed various ML and DL techniques. Table 1 defines a comparative analysis of the researches done in the past in this field. 1. Navulur and Prasad [14] Proposed AGRIPI, a solar-powered IoT system with wireless sensors for real-time agricultural management, enhancing decision-making and resource management. Kim et al. [15] viewed IoT applications in agricultural automation, analyzing sensor technologies and communication methods and discussing their limitations and future potential in Korean agriculture. Roy et al. [16] explored smart agricul-

**Table 1** Comparative analysis of various works done in the field of smart agriculture

| S. No. | Author (Year) | Model | Techniques used |
|---|---|---|---|
| 1 | Giri et al. (2016) [10] | IoT | Agricultural automation, IoT |
| 2 | Wolfert et al. (2017) [19] | Big data | Big data analytics, smart farming |
| 3 | Pivoto et al. (2018) [20] | Smart farming | Smart farming technologies, Brazil |
| 4 | Bacco et al. (2019) [21] | Digitization | Digitization of agriculture, smart farming |
| 5 | Jose et al. (2021) [22] | Artificial intelligence | AI techniques, agriculture revolution |
| 6 | Dhanaraju et al. (2022) [1] | IoT-based systems | Wireless sensors, IoT, cloud computing |

ture as a sustainable, cost-effective solution for future food demands, highlighting technologies like machine vision and regression modeling for optimizing agricultural practices. Sharma et al. [17] examined farm automation using IoT and edge computing to address high network bandwidth needs, ensuring efficient and reliable real-time agricultural operations. Giri [18] Proposed energy-efficient, location-aware IoT and WSN integration for precision agriculture, focusing on secure localization and routing protocols to monitor crop fields effectively.

## 3 Challenges and Open Issues

Based on the previous research we have seen the urgent necessity to improve the existing technologies in the field of Agritech to develop a better way to perform the various agricultural activities. The main challenges faced in this field are as discussed below and are shown in Fig. 3.

1. Farmers need proper training and education to effectively use agritech solutions. Providing accessible training programs and resources can help farmers leverage technology to its fullest potential.
2. Many advanced agritech technologies come with high initial costs, making them inaccessible to small-scale farmers or those in developing regions. Finding ways to reduce costs or provide affordable solutions is necessary for widespread adoption.
3. Rural areas, where agriculture is prevalent, often lack reliable internet connectivity and infrastructure to support advanced agritech systems. Improving infrastructure and ensuring connectivity are crucial for the effective deployment of these technologies.

## 4 Proposed Methodology

This paper presents various algorithms related to crop recommendation from improving the smart agricultural techniques. The classification between the various algorithms is done based on the accuracy percentages. Enhancements were made to improve the accuracy of the existing models. The model was trained on the crop recommendation system dataset, which consists of excel sheets (crop_commendation.csv); this dataset was sourced from kaggle [23]. Despite the diverse sources on Kaggle, the dataset was meticulously scrutinized to ensure its accuracy and reliability through cross-validation with established agricultural data sources. A stratified sampling technique was applied to balance the dataset, ensuring fair representation of all crop categories. Five different image classification models are utilized, and their accuracy is enhanced through the changes in training parameters. The methodology for developing a smart agriculture optimization framework integrating IoT and

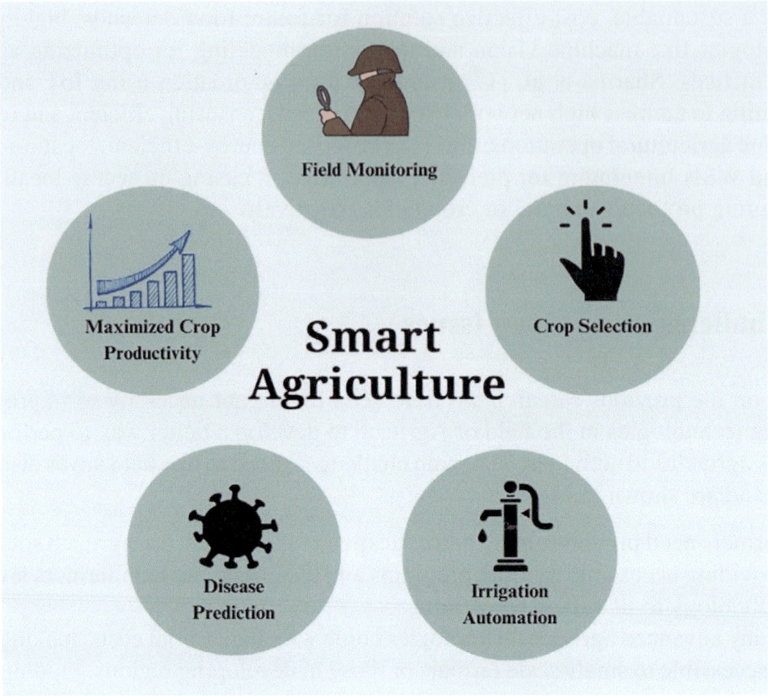


**Fig. 3** Challenges of smart agriculture

machine learning involves several key steps aimed at gathering data, analyzing it effectively, and implementing actionable insights for sustainable crop management.

1. Requirement Analysis: The process begins with a thorough understanding of the needs and goals of a smart agriculture system. This includes the definition of key parameters to be monitored, such as soil moisture, temperature, humidity, crop growth stages, and desired outcomes, such as yield increase, resource efefficiency, and trtransport, it will remain forever.
2. IoT Sensor Deployment: Now, IoT sensors are strategically deployed in fields to collect environmental condition data in real time. The sensors can be embedded in soil, attached to crops, or installed in weather stations to capture a wide range of parameters. The selection and placement of sensors is important to ensure adequate and accurate coverage of the data collected. The deployment strategy considers sensor calibration, network connectivity, and power management to ensure continuous and reliable data acquisition. Sensors are positioned based on a grid-based spatial distribution model, ensuring uniform coverage and minimizing blind spots in data collection.
3. Data collection and integration: IoT sensor data is sent to a centralized system, integrating various sources like soil sensors, weather stations, and satellite imagery for a comprehensive agricultural overview. Data transmission follows a

**Table 2** Sample data from the dataset used in the model

| N | P | K | Temperature | Humidity | pH | Rainfall | Label |
|---|---|---|---|---|---|---|---|
| 90 | 42 | 43 | 20.879744 | 82.002744 | 6.502985 | 202.935536 | rice |
| 85 | 58 | 41 | 21.770462 | 80.319644 | 7.038096 | 226.655537 | rice |
| 60 | 55 | 44 | 23.004459 | 82.320763 | 7.840207 | 263.964248 | rice |

structured protocol, using MQTT and HTTP communication standards to ensure seamless integration with cloud platforms [24, 25]. Time-synchronized data logging ensures consistency, while automated anomaly detection flags potential sensor malfunctions.

4. Data preprocessing: Before analysis, the raw data are preprocessed to remove noise, correct errors, and standardize the format. This can include filtering outliers, interpolating missing values, and normalizing data between sensors to ensure accuracy and precision. Table 2 shows the samples of data chosen from the crop recommendation dataset. Feature engineering techniques, such as principal component analysis (PCA) and correlation-based feature selection, are applied to optimize the dataset for ML modeling. Missing values are handled using KNN imputation, and normalization techniques like Min-Max scaling are used to maintain consistency across diverse sensor data.
5. Machine learning model development: Machine learning algorithms are then trained on the pre-processed data to identify patterns, correlations and enhancements. Supervised learning methods can be used to predict crop yields, detect disease outbreaks, and improve resource quality, while unsupervised learning methods can classify sites based on such characteristics only for targeted interventions. The basic ML model workflow of the proposed model architecture is shown in Fig. 4, providing a clear visualization of the steps involved in the methodology. Model performance is evaluated using multiple metrics, including accuracy, precision, recall, and F1-score, to provide a more comprehensive assessment beyond accuracy alone. Cross-validation techniques such as k-fold validation ($k = 10$) are employed to ensure robustness in model generalization.
6. Application and Testing: Apply and rigorously test the system in real-world agricultural conditions, validating predictions, evaluating recommendations, and measuring impacts on productivity, resource efficiency, and sustainability. This step ensures that the developed IoT-based Agritech system is effective and practical for real-world applications, highlighting its potential to improve agricultural outcomes.
7. Iterative Improvement: Continuously collect feedback from farmers and stakeholders to refine smart agriculture systems, including improving ML models, adjusting sensor placement, and updating algorithms based on environmental changes and agricultural practices.

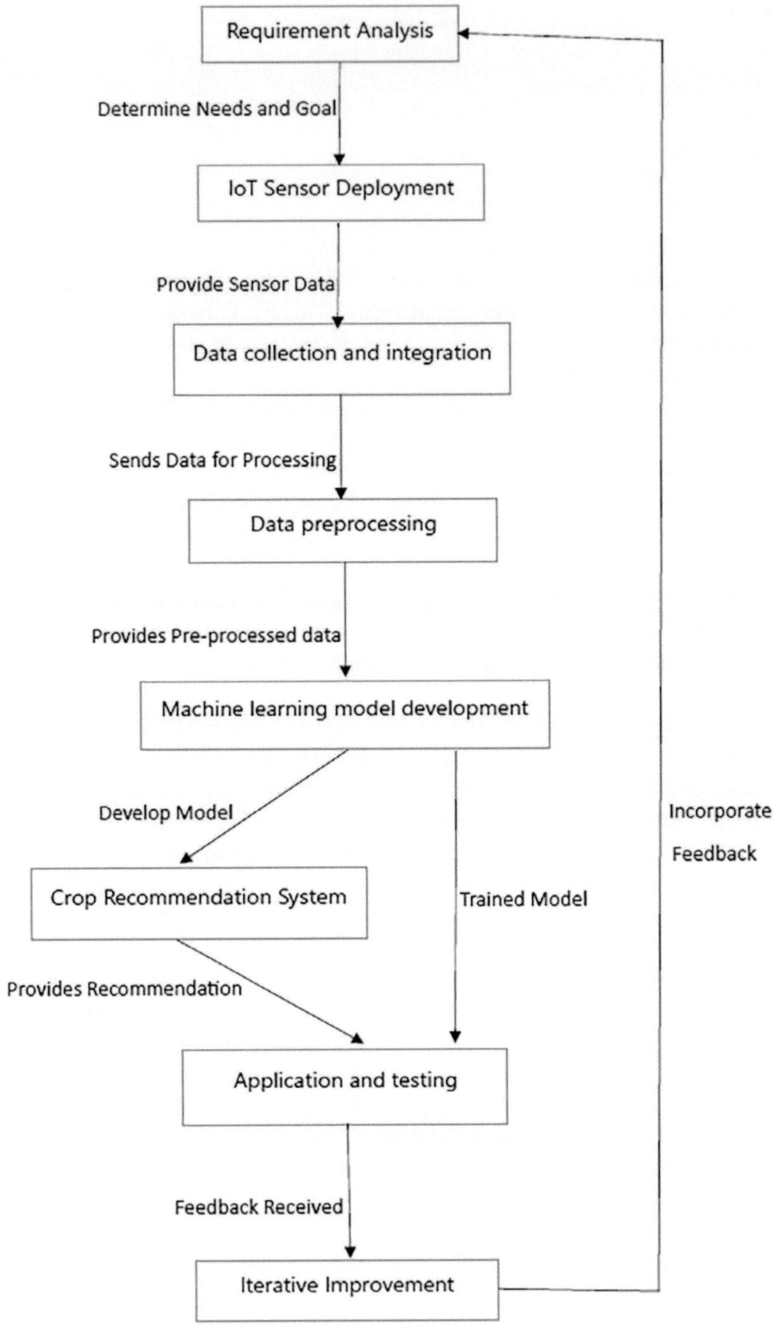


**Fig. 4** Proposed model architecture

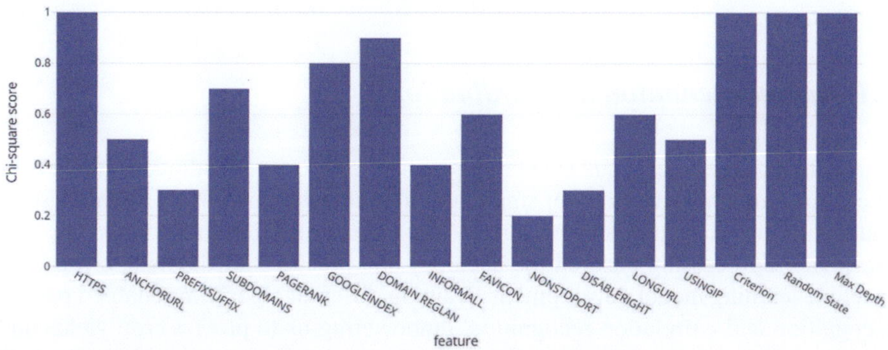


**Fig. 5** Feature analysis and chi-square score graph

After performing these steps, now we come up to explain the results and do the comparative analysis. The different features used in our study and their analysis is shown in Fig. 5.

In smart agriculture, various classification algorithms are utilized to enhance decision-making processes such as crop recommendation and disease detection. Decision Trees (DT) provide interpretable models with around 90% accuracy, though they may overfit complex datasets. Support Vector Machines (SVM) achieve high accuracy, up to 97%, but require careful kernel and parameter selection. Naive Bayes (NB), though simple and based on feature independence, is efficient and effective, often reaching 95% accuracy. Logistic Regression (LR), a linear model, offers ease of interpretation and can achieve 99% accuracy if data relationships are linear. Random Forests (RF), an ensemble method, combine multiple trees to enhance robustness and accuracy, also around 99%. K-Nearest Neighbors (KNN), based on instance similarity, is effective for irregular decision boundaries but can be computationally expensive, with 95% accuracy potential. Neural Networks (NN) are highly flexible and powerful, especially with large datasets, achieving around 98% accuracy, though they require significant computational resources [26]. Lastly, Gradient Boosting Machines (GBM) sequentially correct errors of weak learners, offering around 97% accuracy with effective hyperparameter tuning. Each algorithm's performance varies based on dataset characteristics and complexity [27]. The methodological rigor of this study is ensured through a comprehensive research design that includes extensive requirement analysis, strategic IoT sensor deployment, thorough data collection and integration, meticulous data preprocessing, and robust machine learning model development. Additionally, rigorous application and testing in real-world conditions, followed by iterative improvements based on stakeholder feedback, highlight the study's commitment to practical applicability and continuous refinement. This detailed and systematic approach underscores the appropriateness and robustness of the research methodology.

# 5 Results

## 5.1 Experimentation and Evaluation

Through our experiments and evaluations using the described methodology, we conducted a comprehensive study of smart agricultural techniques. We have laid a solid foundation by using IoT sensors for real-time data collection and connecting data streams. Preprocessing ensured the accuracy and precision of the data, critical for machine learning model development. Training different algorithms enabled pattern recognition and correlation recognition, empowering us to predict crop yields and optimize resource allocation [28]. Each algorithm's implementation within the IoT-based system was carefully assessed. For example, Random Forest and Naive Bayes algorithms, which achieved the highest accuracy scores, demonstrated significant improvements in resource management and crop yield when applied in real-world agricultural settings. In implementation in an agricultural setting, our system demonstrated significant performance improvements, validated through rigorous testing in real-world settings. The dataset was trained on five models Training parameters such as gamma, random_state, criterion, max_depth, etc.

## 5.2 Comparative Analysis

Various smart agriculture algorithms were evaluated, focusing on accuracy. Machine learning models were tested on a crop recommendation dataset from Kaggle in a Linux environment [29]. The impact of adjusting training parameters [30] on image classification accuracy was analyzed, and an IoT-ML-integrated approach was compared with traditional farming. Results showed improved prediction accuracy, resource efficiency, and sustainability. Figure 6 illustrates the heat map of the crop recommendation, while Fig. 7 presents the comparison of classification algorithms based on accuracy.

The comparison of the 8 classification algorithms based on their accuracy is shown in Table 3, with a corresponding graphical representation in Fig. 8. To enhance clarity, a bar chart or line graph as shown in Fig. 7, is recommended instead of the current format, as it would provide a more intuitive visual comparison of algorithm performance. Such a representation would make it easier to distinguish variations in accuracy across different models.

# 6 Conclusion and Future Scope

IoT-based Agritech frameworks optimize agribusiness and crop recommendations using machine learning, enabling real-time monitoring of soil, climate, and crop

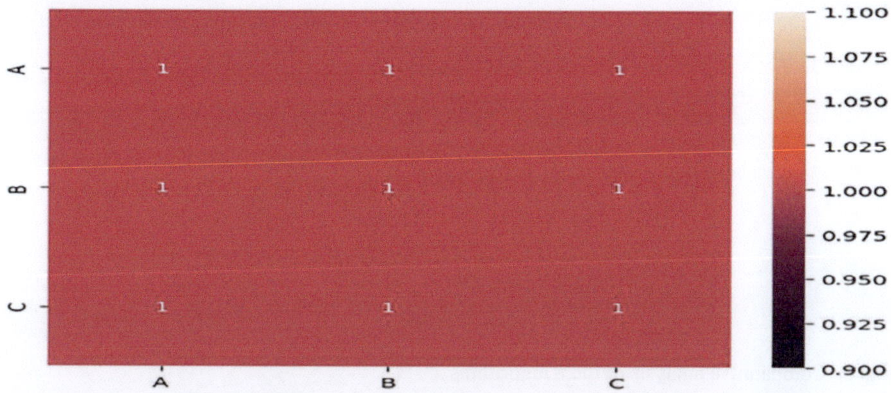


**Fig. 6** Heat map of the crop recommendation indicating what crop to grow

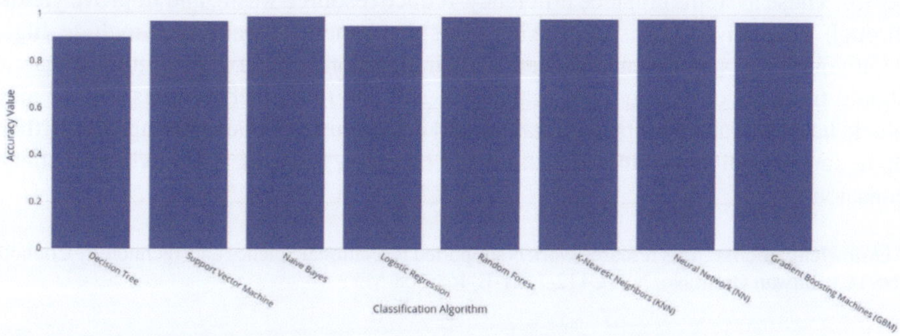


**Fig. 7** Bar chart for representation of the different classification algorithms

**Table 3** Comparison of accuracy values of different classification algorithms

| Classification algorithm | Accuracy value |
|---|---|
| Decision Tree | 0.90 |
| Support Vector Machine | 0.97 |
| Naive Bayes | 0.99 |
| Logistic Regression | 0.95 |
| Random Forest | 0.99 |
| K-Nearest Neighbors (KNN) | 0.98 |
| Neural Network (NN) | 0.98 |
| Gradient Boosting Machines (GBM) | 0.97 |

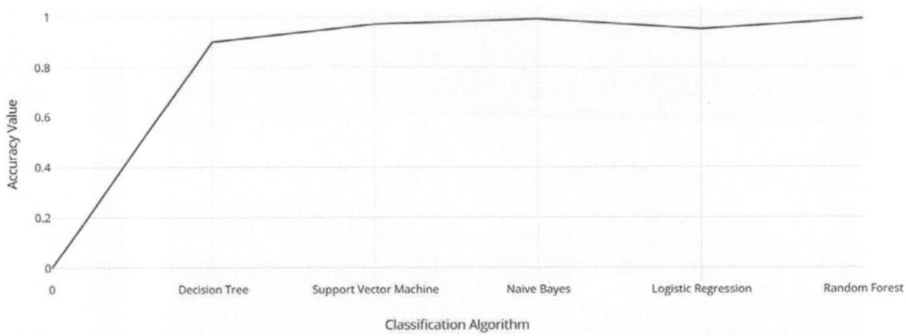


**Fig. 8** Comparative analysis of the 8 algorithms

health. These systems enhance efficiency, reduce resource waste, and improve yields through data-driven decision-making. Despite benefits, challenges include high deployment costs, maintenance needs, and integration complexities. Future research should focus on cost reduction, scalability, and emerging technologies like AI and blockchain. Advancements in climate-smart agriculture and data sharing will further drive sustainability, making farming more efficient, resilient, and environmentally conscious.

**Acknowledgements** This research work is supported by National Science and Technology Council (NSTC), Taiwan Grant No. NSTC112-2221-E-468-008-MY3.

## References

1. Dhanaraju, M., Chenniappan, P., Ramalingam, K., Pazhanivelan, S., Kaliaperumal, R.: Smart farming: Internet of Things (IoT)-based sustainable agriculture. Agriculture **12**(10), 1745 (2022)
2. Sharma, A., Singh, S.K., Kumar, S., Thakur, R., Gupta, B.B., Arya, V.: IoT-enabled smart farming with Industry 5.0. J. High Speed Netw. **30**(3), 477–496 (2024)
3. Sarin, S., Singh, S.K., Kumar, S., Goyal, S., Gupta, B.B., Arya, V., Alhomoud, A.: Enhancing smart grid reliability through cross-domain optimization of IoT sensor placement and communication links. Telecommun. Syst. **88**(1), 1–16 (2025)
4. Thakur, U., Singh, S. K., Kumar, S., Singh, A., Arya, V., Chui, K.T.: Multi-modal sensor fusion with CRNNS for robust object detection and simultaneous localization and mapping (SLAM) in Agile Industrial Drones. In: AI Developments for Industrial Robotics and Intelligent Drones, pp. 285–304. IGI Global Scientific Publishing (2025)
5. Katna, S., Singh, S.K., Kumar, S., Manro, D., Chhabra, A., Sharma, S.K.: Communication systems for drone swarms and remote operations. In: AI Developments for Industrial Robotics and Intelligent Drones, pp. 115–136. IGI Global Scientific Publishing (2025)
6. Garg, S., Kumar, S., Singh, S.K., Kumar, S., Arya, V., Gupta, B.B.: Gestational diabetes prediction using machine learning for consumer electronics healthcare. In: 2024 IEEE International Symposium on Consumer Technology (ISCT), pp. 243–249. IEEE (2024)

7. Arora, L., Singh, S.K., Kumar, S., Gupta, H., Alhalabi, W., Arya, V., Gupta, B.B.: Ensemble deep learning and EfficientNet for accurate diagnosis of diabetic retinopathy. Sci. Rep. **14**(1), 30554 (2024)
8. Kumar, S., Singh, S.K., Nelson, L.: Computational intelligence in decision support: Scope and techniques. In: Uncertainty in Computational Intelligence-Based Decision Making, pp. 219–238. Academic Press (2025)
9. Goyal, S., Kumar, S., Singh, S.K., Gupta, B.B., Arya, V., Chui, K.T.: intelligent task offloading in iot-driven digital twin systems via hybrid federated and reinforcement learning. In: 2024 IEEE Cyber Science and Technology Congress (CyberSciTech), pp 400–405. IEEE (2024)
10. Giri, A., Dutta, S., Neogy, S.: Enabling agricultural automation to optimize utilization of water, fertilizer and insecticides by implementing Internet of Things (IoT). In: 2016 International Conference on Information Technology (InCITe)-The Next Generation IT Summit on the Theme-Internet of Things: Connect your Worlds, pp. 125–131. IEEE (2016)
11. Zhao, G., Li, X., Li, H.: A trusted authentication scheme using semantic LSTM and blockchain in IoT access control system. Int. J. Semant. Web. Inf. Syst. (IJSWIS) **20**(1), 1–27 (2024)
12. Wu, Y., Chen, J.: A lightweight real-time system for object detection in enterprise information systems for frequency-based feature separation. Int. J. Semant. Web. Inf. Syst. (IJSWIS) **19**(1), 1–18 (2023)
13. Srinivasan, C.S.K.: Smart farming-challenges and their solution on agriculture using IoT. Ann. Romanian Soc. Cell Biol. 3983–3996 (2021)
14. Navulur, S., Prasad, M.G.: Agricultural management through wireless sensors and internet of things. Int. J. Electr. Comput. Eng. **7**(6), 3492 (2017)
15. Kim, W.S., Lee, W.S., Kim, Y.J.: A review of the applications of the internet of things (IoT) for agricultural automation. J. Biosyst. Eng. **45**(4), 385-400 (2020)
16. Roy, S., Chakraborty, R.S: Low-cost smart irrigation solution for efficient water use and requirement prediction. Comput. Electr. Eng. 125, 110420 (2025)
17. Sharma, A., Singh, S.K., Kumar, S., Thakur, R., Gupta, B.B., Arya, V.: IoTenabled smart farming with Industry 5.0. J. High Speed Netw., **30**(3), 477-496 (2024)
18. Giri, A., Dutta, S., Neogy, S.: Enabling agricultural automation to optimize utilization of water, fertilizer and insecticides by implementing Internet of Things (IoT). In: 2016 International Conference on Information Technology (InCITe)-The Next Generation IT Summit on the Theme-Internet of Things: Connect your Worlds, pp. 125–131. IEEE (2016)
19. Wolfert, S., Ge, L., Verdouw, C., Bogaardt, M.J.: Big data in smart farming—a review. Agric. Syst. **153**, 69–80 (2017)
20. Pivoto, D., Waquil, P.D., Talamini, E., Finocchio, C.P.S., Dalla Corte, V.F., Vargas Mores, G.: Scientific development of smart farming technologies and their application in Brazil. Inf. Process. Agric. **5**(1), 21–32 (2018)
21. Bacco, M., Barsocchi, P., Ferro, E., Gotta, A., Ruggeri, M.: The digitisation of agriculture: a survey of research activities on smart farming. Array **3**, 100009 (2019)
22. Jose, A., Nandagopalan, S., Akana, C.M.V.S.: Artificial Intelligence techniques for agriculture revolution: a survey. Ann. Rom. Soc. Cell Biol. 2580–2597 (2021)
23. Kifle, M.A.: IoT and ML-Based precision agriculture: a case of Rwanda coffee (Doctoral dissertation, University of Rwanda (College of science and Technology)) (2022)
24. Manasrah, A.M., Aldomi, A.A., et al.: An optimized service broker routing policy based on a differential evolution algorithm in a fog/cloud environment. Cluster Comput **22**, 1639–1653 (2019)
25. Kumar, R., Singh, S.K., Lobiyal, D.K., Chui, K.T., Santaniello, D., Rafsanjani, M.K.: A novel decentralized group key management scheme for cloud-based vehicular IoT networks. Int. J. Cloud Appl. Comput. (IJCAC) **12**(1), 1–34 (2022)
26. Gyamfi, E.K., ElSayed, Z., Kropczynski, J., Yakubu, M.A., Elsayed, N.: Agricultural 4.0 leveraging on technological solutions: study for smart farming sector. arXiv preprint arXiv:2401.00814 (2024)
27. Elijah, O., Rahman, T.A., Orikumhi, I., Leow, C.Y., Hindia, M.N.: An overview of Internet of Things (IoT) and data analytics in agriculture: benefits and challenges. IEEE Internet Things J. **5**(5), 3758–3773 (2018)

28. Singh, I., Singh, S.K., Singh, R., Kumar, S.: Efficient loop unrolling factor prediction algorithm using machine learning models. In: 2022 3rd International Conference for Emerging Technology (INCET), pp. 1–8. IEEE (2022)
29. Singh, S.K.: Linux Yourself: Concept and Programming. Chapman and Hall/CRC (2021)
30. Singh, I., Singh, S.K., Singh, R., Kumar, S.: Efficient loop unrolling factor prediction algorithm using machine learning models. In: 2022 3rd International Conference for Emerging Technology (INCET), pp. 1–8. IEEE (2022)

# Research on Remaining Life Prediction of the Reactor CRDM Based on Monitoring Signal Reconstruction and Adaboost-CNN

**Liu Zhi-long, Zeng Zhen, Wang Qin-chao, Huang Yan-ping, Nie Chang-hua, and Xie Tian-zhou**

**Abstract** With continuous development of modern science and technology, nuclear energy has become one of the most important energy sources in the world. The control rod drive mechanism (CRDM) is the only active device to control the operation of nuclear reactors and is the key equipment for reactor power regulation. As CRDM works in high temperature, high pressure and strong vibration environment, its performance degrades during long-term operation. Based on this, this paper proposes a method for predicting the remaining life of CRDM based on reconstruction of monitoring signals and adaptive boosting with convolutional neural network (Adaboost-CNN). Firstly, the complete ensemble empirical mode decomposition with adaptive noise (CEEMDAN) is used to reduce the noise of monitoring signals of CRDM during operation, and the current signal is corrected by historical data to obtain the reconstructed signal. Secondly, a weak prediction model for the remaining life of CRDM is established based on convolutional neural network (CNN). After that, a strong prediction model can be established based on multiple weak prediction models and Adaboost. Finally, the proposed intelligent remaining life prediction method is verified. The experimental results have shown that the proposed method can accurately predict the remaining life of CRDM and is superior to traditional methods.

**Keywords** CRDM · Remaining life prediction · Adaboost · CNN

---

Z. Liu (✉) · Z. Zeng · Q. Wang · Y. Huang · C. Nie · T. Xie
State Key Laboratory of Advanced Nuclear Energy Technology, Nuclear Power Institute of China, Chengdu, China
e-mail: lzluestc@163.com

 
Y. Ma (ed.), *Proceedings of 4th International Conference on Intelligent Systems Design and Engineering Applications*, Lecture Notes in Electrical Engineering 1478,
https://doi.org/10.1007/978-981-95-1802-9_7

## 1 Introduction

With the long-term rapid development of nuclear energy technology, the reliability requirements for nuclear energy equipment are getting higher and higher. As the only active device to the reactor core, CRDM directly affects the reliable operation of the reactor. Therefore, the research on remaining life prediction of CRDM has a great significance. Research on the health management of nuclear power equipment is ongoing, such as References [1–3]. Vibration signal analysis and artificial intelligence technology are commonly used in the filed of equipment health management. This paper will conduct research on the remaining life prediction of the reactor CRDM based on monitoring signal reconstruction, correction and Adaboost-CNN.

As an important signal analysis method, CEEMDAN is widely adopted in time series prediction and fault diagnosis [4–6]. Reference [4] used LSTM combined with CEEMDAN to predict carbon prices; Reference [5] used an improved CEEMDAN method with CNN to select the grounding wires to solve the reliability problem of grid grounding wires; while in Reference [6], a method based on CEEMDAN and an adversarial neural network was proposed to solve the problem of unbalanced fault samples. References [7–9] used neural networks to improve the reliability of equipment during operation, where Reference [7] proposed an improved CNN to achieve the remaining life prediction of lithium-ion batteries; Reference [8] combined CNN, attention mechanisms, and gated neural units to predict the remaining life of rolling bearings; and Reference [9] proposed a new hybrid deep learning model to realize predictive maintenance, which involves technologies such as CNN, LSTM and so on. In Reference [10], a new neural network framework was proposed to achieve remaining life prediction. Reference [11] used Adaboost to form a random feature mapping method to improve the classification performance of models. Reference [12] adopted Adaboost to improve the robustness of the model to outliers.

This paper proposes a method for remaining life prediction of the reactor CRDM that combines CEEMDAN, the signal correction method, CNN and Adaboost to achieve accurate prediction, where CEEMDAN is used to reduce the noise of CRDM vibration signals, the signal correction method is used to correct the time series data of CRDM monitoring signals, CNN is used to establish a weak prediction model for CRDM remaining life, and Adaboost is used to establish a strong prediction model.

# 2 Introduction Noise Reduction and Correction of CRDM Monitoring Signals

## 2.1 Noise Reduction of Monitoring Signals Based on CEEMDAN

The CEEMDAN method is developed on the basis of the empirical mode decomposition (EMD) method. Its function is to decompose the time domain signal into low-frequency fractions (IMFs) with different main frequency components. The general idea of the CEEMDAN decomposition method is as follows: firstly, add white noise and take advantage of EMD to obtain an IMF; secondly, new white noise is added, obtain a new IMF by performing EMD decomposition on the remaining signal of the previous decomposition; then, repeat above steps until the signal is monotonous and cannot be further decomposed. Note that white noise with a certain amplitude is added in the first decomposition, while each subsequent decomposition adds the white noise component after EMD decomposition.

The steps of noise reduction and reconstruction based on CEEMDAN are as follows:

(1) Assuming that $v(n)$ is the vibration signal of CRDM and $s_i$ is white noise with standard normal distribution. Add $s_i$ to the monitoring signal of CRDM to obtain a new signal sequence:

$$v_i(n) = v(n) + \eta s_i(n) \tag{1}$$

where $i = \{1, 2, 3, \ldots, m\}$ and $m$ is the present number of decomposition.

(2) The first IMF is obtained by performing EMD decomposition on the above signal sequence:

$$\overline{\mathrm{IMF}_1[n]} = \frac{1}{m}\sum_{i=1}^{m} \mathrm{IMF}_{1i}n] \tag{2}$$

(3) Calculate the residual component of the signal after the first decomposition:

$$r_1(n) = v(n) - \overline{\mathrm{IMF}_1}(n) \tag{3}$$

(4) Add a new noise sequence $s_{1i}$ to the residue after the first decomposition to obtain a new superposition signal:

$$r_{1i}(n) = r_1(n) + \eta s_{1i}(i) \tag{4}$$

(5) (5) Decompose the above remaining signals by EMD to obtain the second IMF:

$$\overline{\mathrm{IMF}_2[n]} = \frac{1}{m}\sum_{i=1}^{m} \mathrm{IMF}_{2\mathrm{i}}[n] \tag{5}$$

(6) Calculate the residual component of the signal after the second decomposition:

$$r_2(n) = r_1(n) - \overline{\mathrm{IMF}_2}(n) \tag{6}$$

(7) Repeat above steps to obtain remaining IMFs until the remaining signal is a monotonic function and can no longer be decomposed.

Figures 1 and 2 are the time-domain images of the signals at the early and later stages in CRDM operation, while Fig. 3 shows corresponding frequency-domain image. It is difficult to find the characteristics of signal change at the early and later stages in operation from the time-domain diagram of vibration signals. However, it can be seen from the frequency-domain diagram that at the early stage in operation, the energy of vibration signals concentrates on the high-frequency part; while at the later stage, the characteristics of the vibration signal can be summarized as reduced high-frequency energy and increased low-frequency energy. In this way, extracting the low-frequency signal characteristics of CRDM can achieve the purpose of remaining life prediction. The specific noise reduction method is: select low-frequency IMFs to reconstruct the signal after noise reduction.

The IMF components obtained from CEEMDAN decomposition contain different components of the vibration signal, the IMF component whose main component is the low frequency component is selected and reconstructed into a new signal, which will be the input signal to the intelligent prediction model.

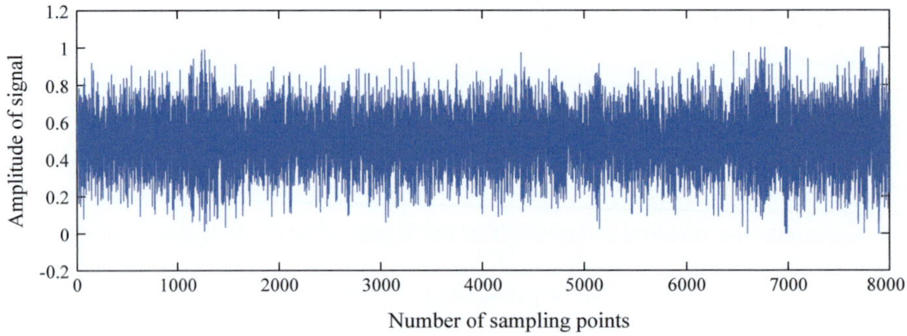


**Fig. 1** Normalized time-domain image of vibration signals at the early stage in operation

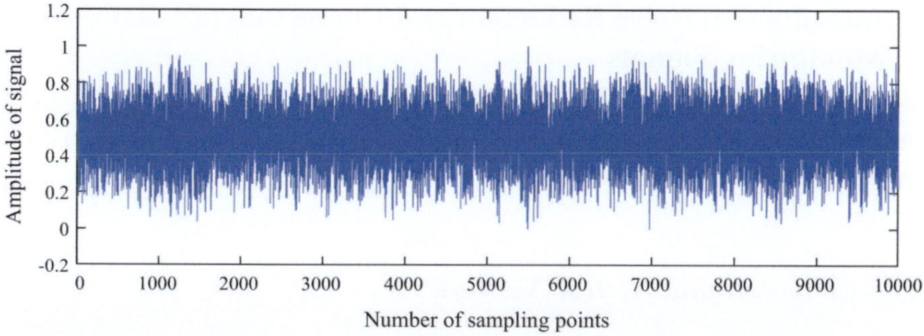


**Fig. 2** Normalized time-domain image of vibration signals at the later stage in operation

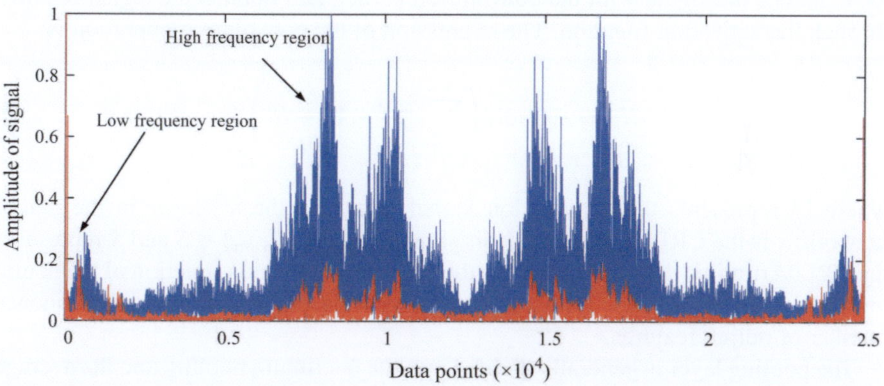


**Fig. 3** Normalized frequency-domain image of vibration signals at the early and later stages in operation

## 2.2 Correction of Denoised CRDM Signals

The degradation characteristics of CRDM in operation concentrate on the low-frequency part of the vibration signal, so the analysis of the frequency characteristics of the signal is the key to predicting the remaining life. Based on the above content, the vibration signal $v_r$ of CRDM after noise reduction and reconstruction is used to correct the current denoised and reconstructed signal so that the remaining life prediction model established later can extract comprehensive and effective degradation characteristics.

The corrected denoised and reconstructed signal is as follows:

$$v_r(n) = \sum_{i=1}^{n} v_r(i) \tag{7}$$

## 3 Introduction Noise Reduction and Correction of CRDM Monitoring Signals

The new model named Adaboost-CNN which combines CNN and Adaboost can accurately predict the remaining life of CRDM.

### 3.1 Convolutional Neural Network

CNN mainly consists of an input layer, a convolutional layer, a pooling layer, a fully connected layer and an output layer, where the convolutional layer convolves the local area of the signal with the convolution kernel, and obtains the signal features through the activation function. The expression of the convolution operation is:

$$X_j^u = \mathrm{sigmoid}\left(\sum_{i=1}^{n} x_i^{u-1} * k_{ij}^u + b_j^u\right) \tag{8}$$

where $k_{ij}^u$ represents the convolution kernel; $u$ means the $u^{th}$ layer in the neural network; $x$ is the CRDM target analysis signal; $\{x_1, x_2, x_3, x_\mathrm{n}\} \in S$ and $S$ represents the CRDM runtime dataset; sigmoid is the nonlinear activation function of the neural network; $i$ is the sequence number of the target analysis signal; and $j$ is sequence number of output features.

The pooling layer is generally divided into the maximum pooling and the average pooling, whose function is to retain the main features of the signal and reduce the amount of calculation to prevent possible over-fitting phenomenon. This paper selects the maximum pooling, its expression is as follows:

$$X_u^j = \mathrm{down}(X_{u-1}^i) \tag{9}$$

where $u$ represents the number of layers; $i$ and $j$ are sequence numbers of input and output, respectively; down means the maximum pooling function.

The fully connected layer further extracts and fuses the features of the output of the pooling layer, and inputs the results to the fitting part for sub-fitting to obtain the output results.

The IMF components whose main components are low frequency components are selected and reconstructed into a new signal, which is then Fast Fourier Transform (FFT) and the transformed signal is fed into the CNN model. The CNN model performs convolutional feature extraction on the reconstructed signal frequency domain waveform and performs lifetime fitting.

## 3.2 Adaboost

Adaboost is an iterative algorithm used to improve the performance of machine learning models. Specifically, for the weak models established by machine learning algorithms, the weights of data samples and each weak machine learning model are iteratively corrected, so as to improve the performance of the weak model and reduce its errors. Modifying the weights of data sample can make the next training of weak models focus on samples with large errors; modifying the weights of several weak model samples can make the weights of models with large errors smaller and those with small errors larger, and finally improve the performance of prediction. CRDM works in a complex environment, considering that vibration signals are featured by high background noise and non-stationary characteristics, it is difficult for traditional machine learning models to extract the features of vibration signal effectively and predict the remaining life of CRDM accurately when performing prediction tasks. In this paper, the remaining life prediction model is established by corrected CRDM monitoring data after noise reduction as well as CNN, and Adaboost is adopted to improve the performance of the established model to form a strong prediction model.

The optimizing process of of Adaboos is as follows: use the established CEEMDAN-CNN model as the basic prediction model and optimize it iteratively. The specific steps are as follows:

(1) Initialize the weights of CRDM historical data samples and assign the same weight value to each sample, as follows:

$$w_n(i) = \frac{1}{n}, i = 1, 2, ..., m, \tag{10}$$

where $m$ is the number of samples, $n$ represents the $n^{th}$ prediction model.

(2) Calculate the fitting error of the CEEMDAN-CNN model:

$$e_n = \sum_{i=1}^{m} w_n(i) \times E_n(i) \tag{11}$$

where $E_n(i)$ is the relative sample error.

(3) Calculate the weight coefficients of the CEEMDAN-CNN model:

$$\alpha_n = \ln \frac{1 - e_n}{e_n} \tag{12}$$

(4) Calculate and update the weight coefficients of each data sample of CRDM:

$$w_{n+1}(i) = \frac{w_n(i)\alpha_n^{1-E_n(i)}}{\sum_{i=1}^{m} w_n(i)\alpha_n^{1-E_n(i)}} \tag{13}$$

(5) Establish a strong prediction model. Combine each weak prediction model after training into a strong prediction model, which can be expressed as follows:

$$c = \arg\max \sum_{n=1}^{N} (\alpha_n \times T_n) \tag{14}$$

where $T_n$ means the $n^{th}$ model.

## 3.3 Residual Life Prediction Method and Model Performance Evaluation

Considering the methods and principles mentioned above, the overall block diagram of the CEEMDAN-Adaboost remaining life prediction model for CRDM can be obtained, as shown in Fig. 4. It is mainly divided into three parts: firstly, CEEDMAN is used to denoise and reconstruct the CRDM vibration signal; secondly, the reconstructed data is corrected; after that, a weak prediction model is established by CNN; finally, Adaboost is used to combine weak models to form a strong prediction model for the remaining life of CRDM.

In order to assess the effectiveness and superiority of the method proposed in this paper, the mean absolute error (MAE) and root mean square error (RMSE) are calculated for performance evaluation.

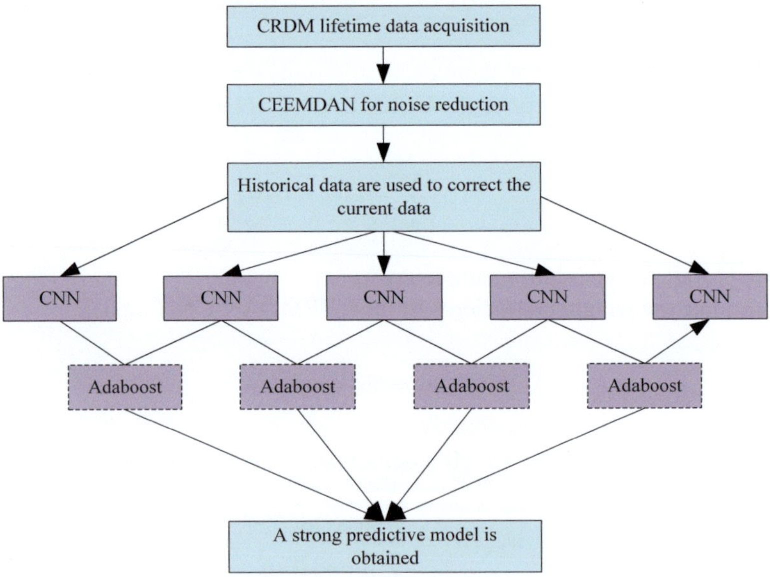


**Fig. 4** The overall block diagram of the remaining life prediction model

$$\mathrm{MAE_{CRDM}} = \frac{1}{n}\sum_{i=1}^{n}|y_i - \hat{y}_i| \tag{15}$$

$$\mathrm{RMSE_{CRDM}} = \sqrt{\frac{1}{n}\sum_{i=1}^{n}|y_i - \hat{y}_i|^2} \tag{16}$$

where $y_i$ is the predicted value and $\hat{y}_i$ is the actual value.

## 4 Experimental Results and Analysis

The experimental data used in this paper comes from the CRDM reliability experiment. The vibration sensor used here is PCB 357B61. Figure 5 shows a charge amplifier named UV-16 for CRDM vibration signal measurement, which can input 8 vibration signals at the same time. Figure 6 shows a CRDM experimental data acquisition system, which mainly includes a data acquisition case and a designed monitoring software for the upper computer. In the experiment, the sampling frequency is 10000 Hz, and a vibration signal is collected to form a data sample every 90 m of the drive mechanism.

Based on the CRDM experimental data above, the proposed remaining life prediction method is verified and compared with other methods.

Figures 7 and 8 show the training curve of the weak prediction model of CRDM remaining life and the curve of the loss function during training process, where data1 to data5 represent the results of the 1st to 5th weak prediction models, respectively. It can be seen from Fig. 7 that the RMSEs in the final training of the five weak prediction models are all close to 0. In Fig. 8, values of the loss functions of the five weak prediction models almost reach 0.

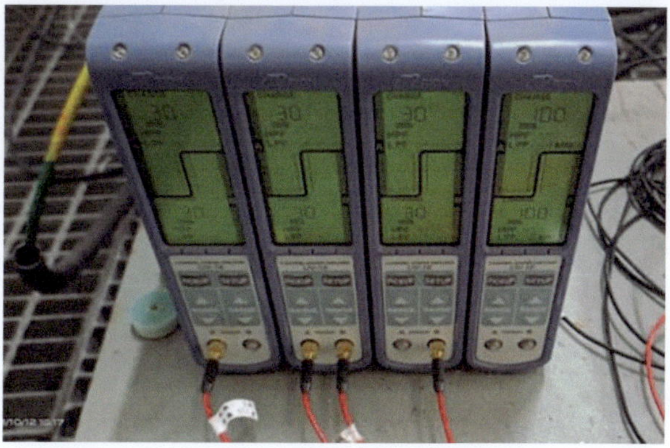

**Fig. 5** The charge amplifier for CRDM vibration signal measurement

**Fig. 6** The CRDM experimental data acquisition system

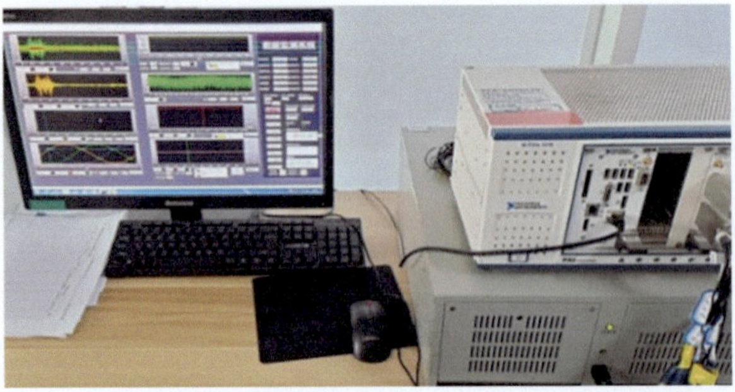

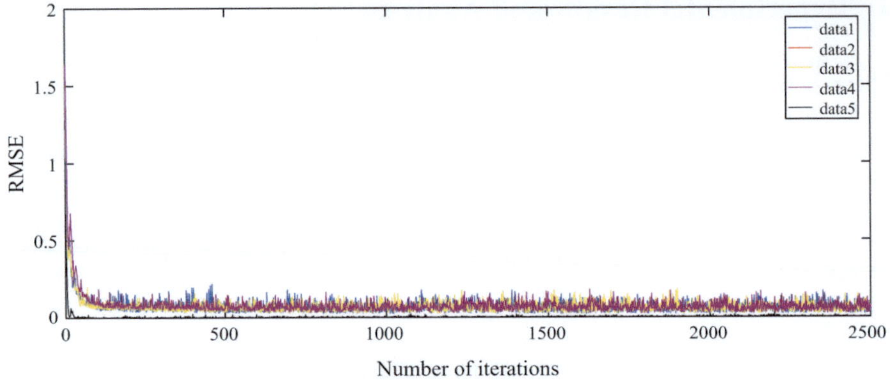


**Fig. 7** The value of RMSEs during training process

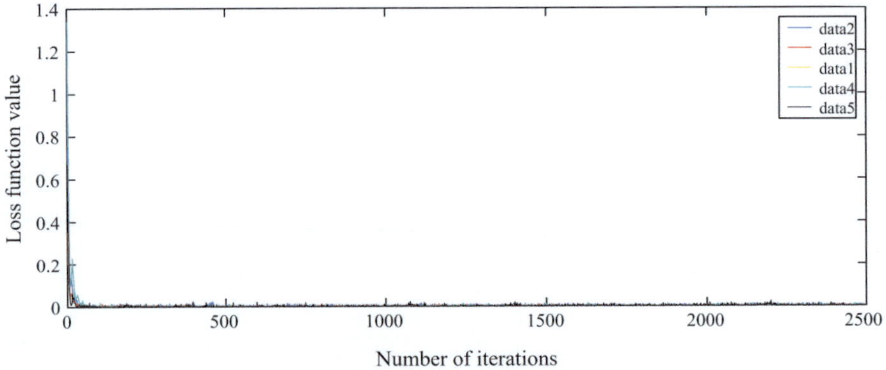


**Fig. 8** The loss function during training process

Figure 9 shows the prediction results of the strong prediction model for the remaining life of CRDM as well as the average results of the weak prediction models, where data1 represents the averaged absolute values of the errors of strong prediction models, while data2 represents the absolute values of the errors of the weak prediction model. When the remaining life of CRDM is 2520 m, the prediction is performed every 90 m. It can be seen that data2 is significantly smaller than data1, that is, the strong prediction model is obviously better than the weak prediction model. At the same time, it proves that the prediction model proposed in this paper can effectively predict the remaining life of CRDM. It is easy to know that the average error of the remaining life prediction is 94%.

The MAE and RMSE values of the proposed method are further calculated and compared with the FFT-CNN, NC-CNN methods, and the results are shown in Table 1 and Fig. 10, NC means noise reduction and correction. It can be seen that the proposed NC-CNN-Adaboost method outperforms the remaining two methods in terms of both the MAE and RMSE metrics. This is due to the fact that the NC-CNN-Adaboost method is able to extract more comprehensive CRDM degradation features and further improve the performance of NC-CNN.

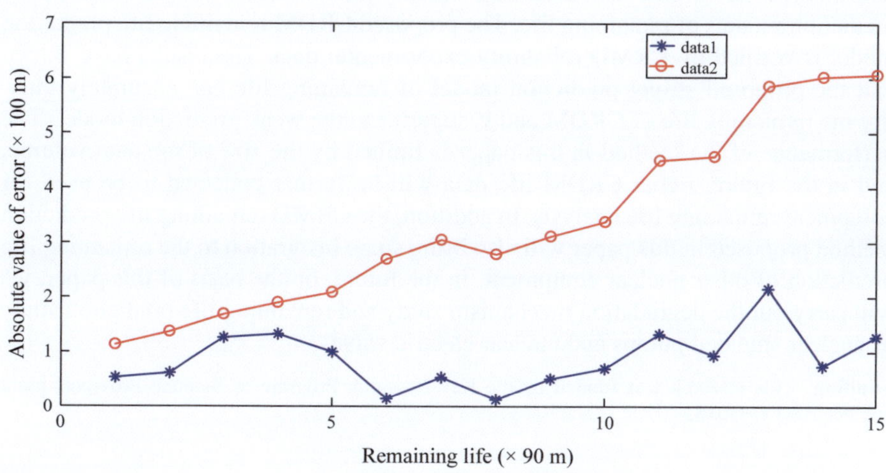


**Fig. 9** The absolute values of the errors of remaining life predictions

**Table 1** Performance of each method

| Method | MAE | RMSE |
|---|---|---|
| NC-CNN-Adaboost | 0.89 | 1.01 |
| FFT-CNN | 3.57 | 3.95 |
| NC-CNN | 3.35 | 3.72 |

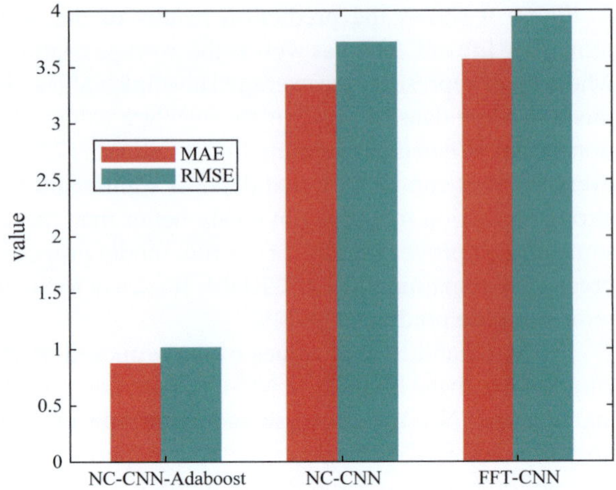


**Fig. 10** Comparison of the performance of each method

# 5 Conclusion

In this paper, CEEMDAN, signal correction method, CNN, Adaboost and other methods are used to achieve: noise reduction of CRDM vibration signals, correction of time series data of monitoring signals, establishment of weak and strong prediction models of remaining life. The proposed CRDM remaining life prediction model is verified by CRMD reliability experimental data. The results have shown that the proposed strong prediction model of remaining life can accurately identify the remaining life of CRDM and is superior to the weak prediction model. The performance of the method in this paper is limited by the size of the data volume, and in the future, richer CRDM life data will be further collected to be used for equipment remaining life analysis. In addition, the CRMD remaining life prediction method proposed in this paper will also bring some inspiration to the remaining life prediction of other nuclear equipment. In the future, on the basis of this paper, we will carry out the degradation mechanism study and remaining life prediction study of nuclear shielded pumps and nuclear electric valves.

**Funding** This research was funded by the Key Research Program of Sichuan Province, grant number 2023YFG0185.

## References

1. Liu, Z., et al.: Noise reduction method for the vibration signal of reactor CRDM based on CEEMDAACN-SK. Electronics **12**(22), 4681 (2023)
2. Liu Z., et al.: Health state identification method of nuclear power main circulating pump based on EEMD and OQGA-SVM. Electronics **12**(2), 410 (2023)

3. Liu, Z., et al.: Intelligent fault diagnosis of nuclear grade electric equipment based on quantum genetic support vector machine. In: 2023 IEEE 12th Data Driven Control and Learning Systems Conference *(DDCLS)*. IEEE (2023)
4. Zhou, F., Huang, Z., Zhang, C.: Carbon price forecasting based on CEEMDAN and LSTM. Appl. Energy **311**, 118601 (2022)
5. Cheng, X.R., Cui, B.J., Hou, S.Z.: Fault line selection of distribution network based on modified CEEMDAN and GoogLeNet neural network. IEEE Sens. J. **22**(13), 13346–13364 (2022)
6. Hou, S.-Z., et al.: Deep-learning-based fault type identification using modified CEEMDAN and image augmentation in distribution power grid. IEEE Sens. J. **22**(2), 1583–1596 (2021)
7. Ma, Y., et al.: A novel method for remaining useful life of solid-state lithium-ion battery based on improved CNN and health indicators derivation. Mech. Syst. Sig. Proc. **220**(2024), 111646.
8. Ma, P., et al.: Prediction of remaining useful life of rolling bearings based on multiscale efficient channel attention CNN and bidirectional GRU. IEEE Trans. Instrum. Meas. **73**, 1–13 (2024)
9. Shoorkand, H.D., Nourelfath, M., Hajji, A.: A hybrid CNN-LSTM model for joint optimization of production and imperfect predictive maintenance planning. Reliab. Eng. Syst. Safety **241**, 109707 (2024)
10. Lee, J., et al.: A machine learning framework for remaining useful lifetime prediction of li-ion batteries using diverse neural networks. Energy and AI **15**, 100319 (2024)
11. Shan, W., et al.: A random feature mapping method based on the AdaBoost algorithm and results fusion for enhancing classification performance. Expert Syst. Appl. **256**, 124902 (2024)
12. Xing, H.-J., Liu, W.-T., Wang, X.-Z.: Bounded exponential loss function based AdaBoost ensemble of OCSVMs. Pattern Recogn. **148**, 110191 (2024)

# Model-Based Prognosis for Hybrid Systems: A Perspective of Patent

**Tsai Hong-Wen**

**Abstract** This article presents a perspective on model-based prognosis for hybrid systems through the lens of patent analysis. The document begins by summarizing four scholarly articles that provide a comprehensive overview of fault diagnostics and prognostics (FDP). These articles highlight the increasing complexity of engineered systems and the need for advanced FDP techniques to ensure reliability and safety. Then transitions to discussing the role of patents in driving innovation and knowledge sharing, acknowledging the challenges associated with patent systems. The document proceeds to analyze several patents, categorizing them into five technical features: fault model generation, fault simulation and testing, fault diagnosis, fault prognosis, and system-level integration and optimization. Specific patents are discussed in detail, showcasing innovative approaches to FDP in various domains, including machine anomaly detection, vehicle diagnostics, analog circuit fault injection, industrial process monitoring, and algorithm optimization. The document concludes by emphasizing the importance of patents in advancing FDP technologies and encouraging further research in this crucial field.

**Keywords** Fault diagnostics and prognostics · Hybrid systems · Patent

## 1 Introduction

### *1.1 A Subsection Sample*

The crucial topic of fault diagnostics and prognostics (FDP) in complex systems is discussed. Some articles explore various aspects, including the methodologies for identifying existing and predicting future faults, the challenges posed by complex

H.-W. Tsai (✉)
Graduate Institute of Patent, National Taiwan University of Science and Technology, Taipei, Taiwan
e-mail: hwtsai@ntust.edu.tw

 
Y. Ma (ed.), *Proceedings of 4th International Conference on Intelligent Systems Design and Engineering Applications*, Lecture Notes in Electrical Engineering 1478,
https://doi.org/10.1007/978-981-95-1802-9_8

systems, the integration of FDP with maintenance strategies, and a survey of recent research and applications. Soleimani et al. provide a comprehensive overview of FDP techniques and their application in complex systems. The article emphasizes the increasing trend in research within this field, especially in electrical, electronic, and mechanical engineering fields. It systematically reviews the four primary FDP approaches: physics-based, data-driven, knowledge-based, and hybrid. The article highlights that each approach has its own advantages and disadvantages, and selecting the appropriate method depends on the specific characteristics of the system and the available data [1].

Jinhan et al. focus on constructing health indicators (HIs) for system-level fault prognosis. It explores the limitations of existing model-driven and data-driven approaches for building HIs, especially in complex systems with multiple components. The article proposes a deep spatiotemporal feature extraction model for HI identification and compares its performance with state-of-the-art methods. The study demonstrates the proposed method's effectiveness in identifying the evolution of relevant HIs, reflecting the degradation level of maintainable components, and reducing the time cost of online fault prognosis [2]. Mojtaba et al. offer a broad survey of recent methodologies and applications in FDP. The article defines key terms, including degradation, remaining useful life (RUL), and failure prognosis, and discusses the main challenges associated with FDP systems. It categorizes failure prognosis methods into model-based, data-driven, and knowledge-based approaches, elaborating on each category's advantages and disadvantages. The article emphasizes the potential of hybrid methods for improving prediction accuracy. It also examines prominent application domains, such as batteries, rotating machinery, and aerospace industries [3]. Canh et al. underscore the importance of online, real-time FDP in preventing catastrophic events. The article presents a systematic methodology integrating sensing, data analysis, condition indicator selection, fault diagnosis, and failure prognosis into a single platform. It details a case study of a ball bearing to illustrate the application of the proposed approach and highlights the benefits of predicting RUL in time for corrective action [4].

The information provided in these four sources offers a valuable overview of FDP in complex systems. The increasing complexity of engineered systems demands innovative approaches to FDP, and the articles highlight various methods, their challenges, and potential solutions. The emphasis on hybrid approaches and real-world applications provides valuable insights for future research and development in this critical field. And Yaguo et al. indicate intelligent fault diagnosis (IFD) utilizes machine learning techniques to establish a direct connection between the ever-increasing volume of machine monitoring data and their corresponding health conditions [5]. Zhibin et al. also propose a new taxonomy and conduct a comprehensive review of unsupervised deep transfer learning for intelligent fault diagnosis, categorizing methods by task [6].

Patents have been a catalyst for innovation, incentivizing research and development by granting inventors exclusive rights. This fosters competition, driving down prices and improving product quality. We took fault models or modeling as the keywords in the abstract of patent specifications and collected more than 1000 patents worldwide. Due to resource limitations, we focused on issued patents from the United States, Japan and Taiwan as the objects of discussion. There are 117 patents selected in total. The issued patents are officially reviewed for industrial applicability, novelty and non-obviousness by each country, so there is a certain degree of reference value.

## 2 Discussion for the Selected Issued Patents

The mentioned 117 issued patents cover a wide range of technologies and applications related to the modeling and diagnosis of system failures. These patents emphasize the importance of identifying the root cause of anomalies, predicting potential faults, and optimizing maintenance strategies. Based on a review of these patents, here are five general technical features commonly employed:

(1) Fault Models:

These models represent the behaviour of a system under various fault conditions. They are used to simulate and predict the effects of different faults, enabling the development of effective diagnostic and prognostic algorithms. For example, patent US11899527B2 describes creating fault models for specific component failures in a machine, allowing the system to identify the most likely cause of an operating anomaly.

(2) Data-Driven Analysis:

This approach leverages historical data, real-time sensor readings, and other relevant information to identify patterns and anomalies indicative of potential faults. Patent US11120646B2 outlines a method for updating fault models based on real-world data, enabling the system to learn and adapt to new failure modes.

(3) Simulation and Testing:

Various techniques are employed to simulate fault conditions and evaluate the effectiveness of diagnostic and prognostic algorithms. Patent US10346273B2 focuses on automated analog fault injection, allowing for systematic testing and validation of diagnostic capabilities.

(4) System-Level Integration:

The patents emphasize the need to consider the interactions between different components and subsystems to effectively diagnose and predict faults. Patent US10877470B2 describes the use of an integrated digital twin to model and monitor an industrial facility, considering the interdependencies between various assets and processes.

(5) Optimization and Adaptation:

Techniques are implemented to optimize the performance of diagnostic and prognostic algorithms and adapt to changing system conditions. Patent US8768668B2 discusses a method for optimizing parameters of diagnostic algorithms using a generic objective function, enhancing the accuracy and effectiveness of fault detection.

These general features highlight the increasing sophistication and integration of fault diagnosis and modeling techniques. The patents demonstrate a trend towards data-driven, model-based approaches that consider the complex interactions within systems to provide more accurate and timely fault detection and prediction. Then we will select representative patents and discuss the five technical features mentioned above separately.

## 2.1 Fault Models

US11899527B2 entitled Systems and methods for identifying machine anomaly root cause based on a selected reduced order model and a selected fault model, is directed to machine condition monitoring and failure analysis, and more specifically, to identifying anomalies and causes of those anomalies. It indicates a method for identifying a cause of a machine operating anomaly including creating a reduced order model (ROMs) for a digital twin model of a selected machine type and feeding current data from a deployed machine into the ROM. This patent filed by Caterpillar Inc. at 2021, also filed at four areas including WO (WIPO), CA (Canada), AU (Austria), and EP (Europe). Figure 1 illustrates a system 400 for creating a ROM for a digital twin model of a selected machine type 402 (e.g., a dump truck) for detecting machine operating anomalies. The plurality of reduced order models 410 can include one or more of a gradient boosting model, a regression model, a recurrent neural network model, or a long short-term memory model.

This invention patent claims a method for identifying a cause of a machine operating anomaly, the method including the summarized key points:

1. Create a Digital Twin: A highly accurate virtual model of the physical machine is created, calibrated to match its normal behaviour.
2. Reduce Model Complexity: A simplified version of the digital twin, a ROM, is created using deep learning techniques to improve computational efficiency.
3. Monitor Real-time Data: Real-time data from the physical machine (telematics data) is fed into the ROM.
4. Anomaly Detection: The ROM's output is compared to the real-world measurements. If a significant difference is detected, an anomaly is identified.
5. Fault Diagnosis: The anomaly is analyzed by comparing the real-world data to the outputs of various fault models, each representing a specific component failure within the digital twin.

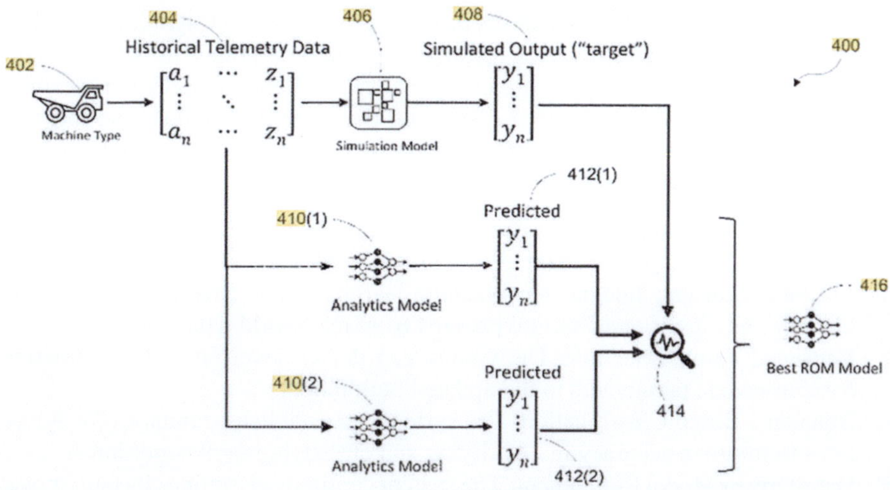


**Fig. 1** Diagram for US11899527B2 illustrating a system for developing a reduced order model for detecting machine anomalies

6. Identify the Cause: The fault model that best matches the real-world data is selected, and the associated component failure is identified as the root cause of the anomaly.

## 2.2 *Data-Driven Analysis*

US11120646B2 is referred to a sophisticated fault diagnostic system, designed to enhance vehicle reliability and maintainability. This system incorporates a dynamic fault model that continuously evolves to accommodate emerging failure modes and their associated symptoms. The future of automotive diagnostics lies in the integration of advanced technologies and intelligent algorithms. The proposed system offers a robust and adaptive solution to ensure optimal vehicle performance and safety. At the core of the system is a comprehensive fault model that encompasses a wide range of potential vehicle failures and their associated symptoms. This knowledge base is continuously updated through a sophisticated machine learning algorithm that analyzes real-world vehicle data to identify emerging failure patterns. By continuously learning and adapting, the system empowers vehicles to proactively identify and address potential issues, enhancing overall reliability and reducing downtime. This proactive approach not only benefits individual drivers but also contributes to a safer and more efficient transportation ecosystem. This patent filed by GM Global Technology Operations LLC at 2019.

Fault models are a cornerstone of modern vehicle diagnostics. By mapping various failure modes to specific symptom combinations, these models enable accurate identification of vehicle malfunctions. For instance, a particular failure mode might be characterized by the simultaneous occurrence of four distinct symptoms. A key innovation presented here involves leveraging machine learning to identify additional, subtle symptoms that can differentiate between failure modes that may exhibit similar initial signs. This enhanced diagnostic precision allows for more targeted troubleshooting and repairs. This invention patent claims the method including the summarized key points:

1. Adaptive Fault Diagnosis System: The system incorporates a dynamic fault model that can learn and adapt to new failure modes and symptoms.
2. Machine Learning Integration: Machine learning techniques are employed to identify and classify new failure patterns from real-world data.
3. Enhanced Fault Detection: The system can detect novel faults by comparing current vehicle parameters to the updated fault model.
4. Boundary-Based Classification: The system can establish boundaries for symptoms to improve accuracy in identifying and classifying fault conditions.
5. Data-Driven Model Refinement: The system continuously refines its fault model by analyzing data and adjusting boundaries.
6. Structured Fault Model: The fault model is organized in a tabular format with rows representing failure modes and columns representing symptoms.

## *2.3 Simulation and Testing*

US10346273B2 is referred to traditional analog fault injection with a manual and time-consuming process. The approach involves creating a comprehensive list of potential faults, including custom fault types tailored to specific circuit vulnerabilities. These faults are then systematically injected into the analog circuit, one at a time, by replacing predefined placeholders with the desired fault models. Advanced simulation techniques are employed to assess the circuit's response to each injected fault, enabling the identification of potential weaknesses and the quantification of fault coverage. By automating this process, we significantly reduce the time and effort required to perform thorough fault injection analysis. This empowers engineers to rigorously test their designs and ensure the highest levels of functional safety and reliability. This patent filed by Analog Devices Global ULC at 2017.

This patent presents a ground-breaking approach to automated analog fault injection, designed to significantly enhance the reliability and safety of complex analog circuits. The proposed method involves several key steps:

1. Fault Model Creation: it meticulously creates a comprehensive list of potential fault models, encompassing a wide range of failure scenarios, including both standard fault types and custom fault models tailored to specific circuit vulnerabilities.
2. Fault Injection: it strategically inserts fault placeholders into the analog circuit at various locations, including critical nodes and sub circuit ports. These placeholders act as injection points for simulating different fault conditions. By

systematically replacing these placeholders with fault models from the list, it can systematically test the circuit's resilience.

3. Fault Simulation and Detection: advanced simulation techniques are employed to assess the circuit's behaviour under each fault condition. It monitor the circuit's response to determine whether the injected fault is detectable by built-in test structures or other diagnostic methods.
4. Results Analysis and Reporting: the detailed report is generated, summarizing the results of the fault injection process. The report provides valuable insights into the circuit's fault coverage, identifying potential weaknesses and areas for improvement.

## *2.4 System-Level Integration*

US10877470B2 is introduced an innovative approach to industrial process monitoring through an integrated facility digital twin (IFDT). This advanced system leverages a combination of static and dynamic models to provide a holistic view of the entire facility, including detailed asset and flow sheet models. By analyzing real-time data and historical trends, the IFDT can identify potential anomalies, predict equipment failures, and optimize operational performance. This unified approach surpasses traditional digital twin solutions by offering enhanced security, simplified management, and improved predictive capabilities. By empowering operators with actionable insights, the IFDT drives significant improvements in efficiency, reliability, and overall productivity. This patent owned by Honeywell International Inc. since 2017, also filed at four areas including WO (WIPO), CN (China), AU (Austria), and EP (Europe).

This patent leverages a sophisticated aggregation algorithm that seamlessly combines static and dynamic models to provide real-time insights into the health and performance of the entire industrial facility. This innovative solution empowers operators to make informed decisions, optimize operations, and proactively address maintenance needs. By enhancing visibility, efficiency, and reliability, the digital twin (DT) drives significant improvements in overall plant performance. The proposed method involves several key steps:

1. Model development: static models can create detailed digital representations of the physical assets and processes, including: asset model, flow sheet model, and dynamic models.
2. Data Integration: to collect real-time data from various sensors and devices across the facility.
3. Model Integration and Aggregation: to combine static and dynamic models to create a comprehensive digital twin.
4. Performance Monitoring and Analysis: to monitor key performance indicators (KPIs) and identify deviations from normal operating conditions.
5. Decision Support and Optimization: to provide actionable insights to operators and engineers.

## 2.5 Optimization and Adaptation

US8768668B2 is introduced a novel approach to optimizing the performance of diagnostic and prognostic algorithms. By systematically tuning the parameters of a diverse set of algorithms, we can significantly improve their accuracy and reliability. The method involves feeding a large dataset of labelled sensor data into each algorithm. The algorithms then generate diagnostic labels for each data point, which are compared to the ground truth labels. A generic objective function is employed to evaluate the performance of each algorithm based on metrics such as accuracy, precision, and recall. To optimize the parameters, we utilize advanced optimization techniques that iteratively adjust the values to minimize the objective function. This iterative process ensures that the algorithms are fine-tuned to the specific characteristics of the system being monitored. By optimizing multiple algorithms simultaneously, we can create a robust and adaptable diagnostic system that can accurately detect and predict faults, leading to improved system reliability and reduced downtime. This patent filed by Honeywell International Inc. at 2012, also filed at CN (China) area. This patent systematically tuning the parameters of a diverse set of algorithms, it can significantly improve their accuracy and reliability. The proposed method involves several key steps:

1. Data Preparation: a comprehensive dataset of labelled sensor data is collected, with each data point associated with a known ground truth label.
2. Algorithm Selection and Parameterization: a diverse set of diagnostic algorithms, such as statistical, machine learning, or knowledge-based techniques, is selected.
3. Diagnostic Label Generation: each algorithm processes the input data and generates a computed diagnostic label for each data point. The computed label can be a specific fault category or a probability distribution over multiple categories.
4. Performance Evaluation: a confusion matrix is constructed to compare the computed labels with the ground truth labels. A weighted objective function is defined to assess the overall performance, considering factors like accuracy, precision, recall, and false positive/negative rates.
5. Parameter Optimization: an optimization algorithm, such as gradient descent or genetic algorithms, is employed to iteratively adjust the parameters of the algorithms. The goal is to minimize the objective function, leading to improved diagnostic accuracy.

# 3 Discussion

The analysis of patents related to model-based prognosis for hybrid systems reveals significant advancements in addressing the challenges of fault diagnostics and prognostics (FDP) in complex engineering systems. Patents like US11899527B2, US11120646B2, US10346273B2, US10877470B2, and US8768668B2 demonstrate innovative approaches to fault model generation, simulation, diagnosis, prognosis, and system-level integration. These innovations, driven by the need for enhanced reliability, safety, and efficiency in critical systems, showcase the potential of model-based prognosis in diverse applications such as machine anomaly detection, vehicle diagnostics, and analog circuit fault injection. The continuous development of sophisticated algorithms, coupled with advancements in sensing and data analytics, is paving the way for more robust, accurate, and adaptable FDP systems. Continued research and development in this field are essential to further improve the performance of these systems and unlock their full potential in real-world scenarios. The insights gleaned from the analysis of patents, alongside the comprehensive overview provided by scholarly articles, provide a valuable foundation for future advancements in FDP technologies.

## References


1. Morteza, S., Felician C., Daniel N.: Diagnostics and prognostics for complex systems: a review of methods and challenges. Qual. Reliabil. Eng. Int. 3746–3778 (2021)
2. Jinhan, Z., Diyin, T., Jinsong, Y., Yue, S., Xuhui, L., Hao, L.: Construction of interpretable health indicators for system-level fault prognosis driven by inductive generative learning. IEEE Trans Instru. and Measure. **73**, 3534612 (2024)
3. Mojtaba, K., Mehrdad, S., Marcos, E.O., Roozbeh, R.F., Khashayar, K.: Failure prognosis and applications—a survey of recent literature. IEEE Trans. Reliabil. **70**, 728 (2021)
4. Canh, L., Kwok, T., Carl, S.B., Romano, P., George, J.V.: Fault diagnosis and failure prognosis for engineering systems: a global perspective. IEEE Conf. Auto. Sci. Eng. **108** (2009)
5. Yaguo, L., Bin, Y., Xinwei, J., Feng, J., Naipeng, L., Asoke, K.N.: Applications of machine learning to machine fault diagnosis: a review and roadmap. Mech. Sys. Sig. Proc. **138**, 106587 (2020)
6. Zhibin, Z., Qiyang, Z., Xiaolei, Y., Chuang, S., Shibin, W., Ruqiang, Y.: Applications of unsupervised deep transfer learning to intelligent fault diagnosis: a survey and comparative study. IEEE Trans. Instru. Measur. **70**, 1–28 (2021)

# From Preservation to Innovation: A Comprehensive Review of Generative AI in the Design Research of Traditional Heritage Furniture

**Wang Yali, Xi Yuchen, Liu Xinxiong, Gan Yan, and Xiang Xingrui**

**Abstract** This study focuses on the innovative potential of generative models in traditional furniture cultural heritage design, emphasizing the integration of profound cultural connotations into modern design practices with the support of advanced intelligent technology. The study uses CiteSpace scientometric analysis and content analysis methods to sort out relevant literature from the Web of Science and China National Knowledge Infrastructure databases from 2000 to 2024, identify key literature, core research teams and important research topics, and dynamically present the application trend of generative models in the field of furniture design through keyword highlighting analysis and hot spot evolution path tracking. The results show that generative models play an important role in promoting the transformation of traditional furniture cultural heritage design from conservative inheritance to modern innovation. International research focuses on applying intelligent means such as diffusion models, multimodal generation technology and augmented reality, while local Chinese research pays more attention to the digital expression and innovative inheritance of traditional cultural symbols. The complementarity of the two points out the direction for the future: integrating the most cutting-edge technology with profound cultural connotations to achieve a deep integration of traditional furniture cultural heritage at the level of cultural protection and design innovation. Research should further optimize algorithm performance, expand high-quality dedicated data sets, develop multimodal intelligent collaborative design tools, and strengthen interdisciplinary cooperation. The above efforts will help inject new impetus into traditional furniture design through generative models based on inheriting cultural heritage and promoting its coordinated development in protection and innovation.

**Keywords** Cultural heritage · Furniture design · Generative models · Cultural preservation and digitalization · Artificial intelligence in design

---

Y. Wang · Y. Xi · X. Liu · Y. Gan (✉) · X. Xiang
Huazhong University of Science and Technology, Wuhan, China
e-mail: ganyan@hust.edu.cn

 
Y. Ma (ed.), *Proceedings of 4th International Conference on Intelligent Systems Design and Engineering Applications*, Lecture Notes in Electrical Engineering 1478,
https://doi.org/10.1007/978-981-95-1802-9_9

# 1 Introduction

With globalization and informatization, traditional furniture cultural heritage faces unprecedented challenges in modernization transformation and sustainable development. With the acceleration of modernization, traditional crafts and cultural symbols with profound historical and cultural connotations have repeatedly collided and impacted the contemporary social value system, making many traditional furniture cultural heritage projects face various difficulties [1]. Take Ming-style furniture as an example. Its concise shape, exquisite craftsmanship, and elegance are regarded as the model and pinnacle of Chinese traditional furniture culture [2]. However, the rapid development of modern society is making traditional furniture gradually lose its appeal in the market. The reason is that the function and aesthetics of traditional furniture can no longer meet the needs of contemporary consumers for personalization, modernization and versatility [3]. In addition, the complexity of traditional craftsmanship, the scarcity of raw materials and the high manufacturing cost also limit the adaptability of traditional furniture and cultural heritage in the modern market. How to inject new vitality into traditional furniture cultural heritage through the combination of innovative design and modern technology while maintaining cultural characteristics has become an important issue that needs to be solved urgently by academia and industry.

Globally, the integration of "cultural heritage and innovation" is gaining increasing attention. The development of intelligent technology has brought new opportunities for the protection and design innovation of traditional furniture. Intelligent design tools represented by generative artificial intelligence technologies such as generative adversarial networks (GAN) and diffusion models can provide strong technical support for pattern extraction, morphological generation and creative design of traditional furniture. These technologies not only help to explore the connotation and potential of cultural symbols and realize the organic integration of tradition and modernity but also open up new paths for the digital protection and market application of traditional furniture. Therefore, the innovative design research on the cultural heritage of traditional furniture will not only help promote cultural protection and inheritance but also promote the practical exploration of the deep integration of traditional culture and modern society, thereby playing a positive role in promoting cultural diversity, developing cultural industries, and enhancing cultural soft power.

However, under the current research background, there are still many unsolved problems in the inheritance and modern application of traditional furniture cultural heritage: First, how to effectively inherit the cultural value of traditional furniture in modern design is still a core challenge; second, at the level of innovative design, the design language and modelling logic of traditional furniture need to be optimized, and it is difficult to break through the existing model; in addition, the application of generative model technology for traditional furniture is still in the exploratory stage, and existing research is primarily a theoretical discussion or single case analysis, and there are still deficiencies in the systematic research and technical practice of specific cultural scenes; finally, the marketization path of traditional furniture cultural heritage is still not clear enough. While maintaining cultural characteristics, improving its market competitiveness through design innovation is a dual issue that designers and researchers urgently need to face.

Based on the above problems, this study will be carried out from the following aspects: First, using bibliometrics and content analysis methods to systematically sort out the research status and hot spots of generative models in the innovative design of traditional furniture cultural heritage, revealing the core path and theoretical basis of its technical application. Secondly, analyze how generative models can extract and reconstruct cultural symbols in traditional furniture-related research objects and provide technical support for the deep integration of culture and technology. Secondly, explore the application scenarios of generative models in traditional furniture design innovation, including form generation, function optimization and personalized design, to tap the modern application potential of traditional furniture cultural heritage. Finally, based on the comprehensive research results, a trend analysis is conducted, and the future development strategy of traditional furniture cultural heritage design is proposed from the perspectives of intelligence, greening and globalization, providing new research perspectives and practical directions for cultural heritage protection and innovative design.

## 2 Research Framework and Methodology

In order to ensure the scientificity and rigour of theoretical basis, data source and research methods, this study first designed a structured literature review framework (see Fig. 1), aiming to deeply analyze the generative design model and its innovative design strategy in the field of traditional furniture cultural heritage design. The research framework comprises five principal components: data collection, data screening, bibliometric analysis, core literature content analysis, and a summary and outlook.

The first step is to obtain literature data through two major international and Chinese academic databases in the data collection stage. The international literature data comes from the Web of Science Core Collection; the Chinese literature data comes from China National Knowledge Infrastructure (CNKI), covering research results related to cultural heritage, furniture design, generative design models and innovative design outside China.

The second step is to screen the collected literature strictly in the data screening stage to ensure its high relevance to the research topic and to set screening criteria such as keywords, document type, collection time, and language. In addition, by screening cross-disciplinary literature, the study ensures the multidimensionality and representativeness of the data.

The third step is to enter the bibliometric analysis stage. This study uses bibliometric tools such as CiteSpace [4] and Nvivo to systematically analyze the selected literature, revealing the research field's historical context and hot-spot evolution.

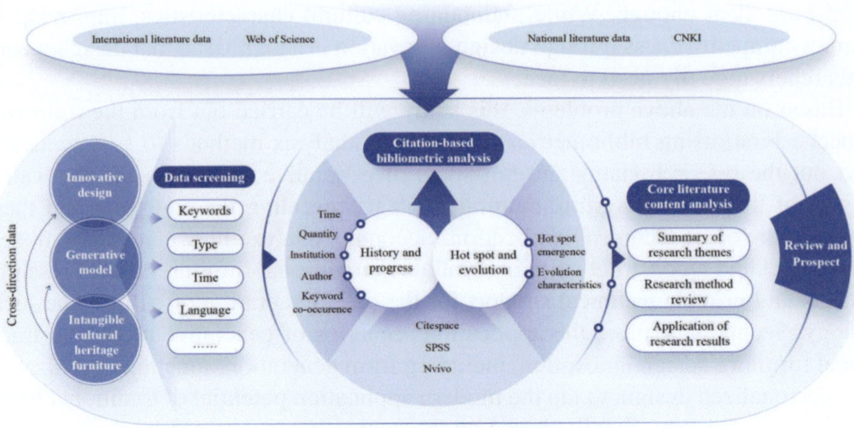


**Fig. 1** Literature review research framework

The fourth stage of the process is to analyse the core literature content. This study employs the content analysis method [5]. The theory and practice of traditional furniture cultural heritage design, including cultural inheritance, generative design models and innovation strategies, are subjected to rigorous examination.

In the fifth step, the research enters the summary and outlook stage. This stage involves a systematic summary of the existing literature based on the above analysis and a proposal of prospects for future research.

## 3 Bibliometric Analysis with CiteSpace

### *3.1 The Specific Operational Steps of Scientometrics*

This study conducted systematic literature retrieval and screening from both the international and Chinese levels to comprehensively analyze the research status of the two major themes, "Innovative Design of traditional Cultural Heritage Furniture" and "Innovative Design of traditional Cultural Heritage Furniture Based on Artificial Intelligence."

In the research direction of traditional cultural heritage furniture and innovative design at the international level, through the Web of Science (WOS) core database, with "Cultural Heritage OR Furniture OR Traditional Craft" and "Innovation Design OR Creative Design" as the subject words, the search time range is from 2000 to 2024, the document type is limited to "Articles", and the language is "English". Finally, 2139 relevant documents were screened out, covering the research trends and technical applications of innovative cultural heritage furniture design worldwide. At the Chinese level, through the China National Knowledge Infrastructure (CNKI) database, with Cultural Heritage + Furniture + Traditional Craft" and "Innovative Design + Creative Design" as the subject words, the time range is also from 2000 to 2024, the document type is limited to "Journal Articles", and the scope is "Beijing Nuclear" + "CSSCI". Finally, 339 relevant documents were obtained, reflecting the latest progress in design innovation and cultural protection research in China's cultural heritage furniture field.

In order to explore the application of artificial intelligence technology in traditional cultural heritage furniture, the study also retrieved literature from both international and Chinese levels. At the international level, through the Web of Science core database, with "Cultural Heritage OR Furniture OR Traditional Craft" and "Artificial Intelligence OR Generative Model OR Machine Learning" as the subject words, the search time range is from 2000 to 2024. Subsequently, 598 documents relating to international issues were deemed unsuitable for inclusion in further consideration. These documents encompassed the utilisation of pioneering technologies, including generative models and deep learning, in conceptualising cultural heritage furniture. At the Chinese level, through the China National Knowledge Infrastructure (CNKI) database, with "cultural heritage + furniture + traditional craft" and "artificial intelligence + generative model + model + deep learning" as the subject words, the search range is the same as above. Finally, 237 relevant documents were obtained, reflecting China's research trend in combining artificial intelligence technology and traditional heritage furniture.

Based on the above screening results, this study uses bibliometric analysis methods. It combines CiteSpace visualization tools to systematically analyze the research context and hot spots in cultural traditional heritage furniture design and its combination with artificial intelligence from the following three aspects: (1) spatiotemporal distribution and achievement statistics. (2) analysis of critical authors and institutions. (3) core themes and hot spot clustering.

## 3.2 The Specific Operational Steps of Scientometrics

**Spatial–temporal Distribution**. This study divides the timeline into three phases to understand the temporal and spatial distribution of research at the intersection of cultural heritage (CH) furniture innovation and artificial intelligence (AI): Initial stage (2000–2009), rapid growth (2010–2019), and deepening development (2020–2024).

*Traditional Heritage Furniture Innovation Research (2000–2024).* The field of Traditional heritage furniture innovation has seen significant growth over time. In the early stages (2000–2009), research was focused on cultural heritage preservation and initial design innovations, with the USA leading in publications. From 2010 to 2019, the field expanded rapidly, with China emerging as the primary contributor, focusing on combining generative models with cultural elements. From 2020 onwards, China has seen dominance, particularly in the application of smart manufacturing, personalized design, and the digitalization of cultural symbols.

*GAI and Traditional Heritage Furniture Research (2000–2024).* The intersection of AI and traditional cultural heritage furniture took shape during the second phase (2010–2019) as AI technologies advanced. China led this phase with applications in generative models and optimizing traditional crafts. In the third phase (2020–2024), research has rapidly progressed, with China maintaining a significant lead, particularly in the use of Generative Adversarial Networks (GANs) and diffusion models in furniture design, while the USA and UK continue to contribute to design optimization and user experience enhancement using AI technologies.

**Collaborative Network**. Internationally, researchers focus on optimization techniques, sustainable material application, and intelligent generative design, while in China, efforts are directed toward digital cultural inheritance and AI-driven intelligent manufacturing.

In the international context, the cross-disciplinary exploration of innovative design and artificial intelligence in cultural heritage furniture primarily concentrates on advancing optimized design tools, sustainable material applications, and intelligent generation technology. Notable scholars in this field include Smardzewski, Gordon Cheng, and others. Prominent research institutions include Mississippi State University and the Technical University of Munich. In China, the research is more oriented towards the digital preservation and representation of traditional cultural heritage symbols and intelligent manufacturing driven by artificial intelligence. Notable scholars include Wu Zhihui and Guan Huiyuan, while prominent institutions include Nanjing Forestry University and the Academy of Fine Arts at Tsinghua University. International and domestic research generally have distinct characteristics regarding key directions and technical applications. Collectively, they facilitate the integrated development of traditional cultural heritage furniture design and intelligent technology.

## 3.3 Knowledge Map and Keyword Clustering

**Keyword clustering in the field of Traditional cultural heritage furniture and innovative design**. According to the WOS keyword clustering diagram (Fig. 2a), the world has formed several representative research themes in the innovative design of Traditional cultural heritage furniture, covering many aspects such as cultural protection, technological innovation and sustainable development. First, "intangible cultural heritage" (#0) is the core research direction in this field, focusing on protecting traditional culture through modern design concepts and promoting the revival of traditional handicrafts and cultural symbols. Secondly, with the development of artificial intelligence technology, the application of "deep learning" (#2) and "intelligent model" (#6) in the design of traditional cultural heritage furniture has gradually increased, combining anthropometry for structural optimization and ergonomic analysis, improving design accuracy and user experience. At the same time, the research on "furniture connection" (#1) focuses on integrating traditional mortise and tenon structure optimization and modern connection technology to improve furniture stability and adapt to modern manufacturing needs. Life cycle assessment has become vital for sustainable design, reducing environmental impact and extending product life by optimizing material selection and design. In addition, "motion recognition" (#7) and "anthropometry" (#4) combined with user behaviour analysis provide data support for improving furniture functionality and comfort, while "conceptual design" promotes the design of innovative furniture products by exploring the modern significance of cultural symbols. Finally, the research on "thermal comfort" (#3) and "coating" (#8) improves the physical properties and environmental adaptability of furniture through material selection and coating technology while focusing on the application of environmentally friendly materials to meet the dual needs of modern design for comfort and sustainability.

As can be seen from the keyword clustering diagram (Fig. 2b), the research on innovative design of Chinese cultural heritage furniture focuses on multiple hot topics. Keywords such as "furniture structure" (#0) and "mortise and tenon structure" (#6) indicate that the research focuses on the modern application of traditional structures, significantly optimizing the mechanical properties and adaptability of mortise and tenon technology through structural simulation and computer-aided design so that it can be integrated into modern manufacturing processes. Keywords such as "traditional craft" (#2) and "intangible cultural heritage" (#3) reflect the importance of traditional cultural inheritance and innovation and endow traditional cultural elements with new aesthetic value and functionality through design methods. "Design innovation" (#4) and "user experience" (#5) emphasize the role of modern design technology in improving furniture innovation and user emotional connection, and tools such as generative design and 3D modelling are widely used. The prominence of keywords such as "hierarchical analysis method" (#7) and "TRIZ theory" (#8) illustrates the significance of design decision-making tools in optimizing traditional cultural heritage furniture design.

Conversely, "eye-tracking technology" (#1) and "digitalization" (#9) emphasize the utilization of digital technology in user behaviour analysis and cultural communication. In addition, "sustainability" (#14) and "product design" (#11) reflect the focus on environmentally friendly materials and green manufacturing processes, emphasizing the minimization of environmental impact and market adaptability during the life cycle. Overall, the research on Chinese cultural heritage furniture design has shifted from traditional cultural protection to technological integration and innovative design, exploring the value continuation of traditional culture in contemporary furniture through modern tools and intelligent technology.

**Keyword clustering in the intersection of traditional cultural heritage furniture and artificial intelligence**. According to the WOS keyword clustering diagram (Fig. 3a), the world has formed several research topics in the intersection of traditional cultural heritage furniture and artificial intelligence, covering artificial intelligence, computer vision, intelligent manufacturing, Internet of Things, statistical analysis and network security. First, technologies such as "generative adversarial network" (GAN) (#2), "deep learning" (#13) and "artificial neural network" (#14) are widely

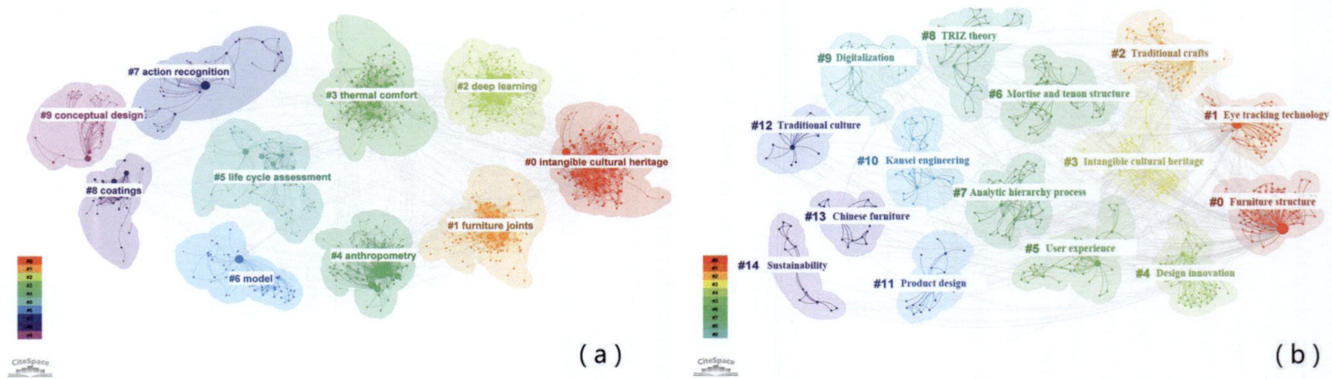


**Fig. 2** **a** Keyword clustering diagram for innovative design research of global traditional cultural heritage furniture. **b** Keyword clustering diagram of innovative design research on traditional cultural heritage furniture in China

used in traditional cultural heritage furniture design. The design accuracy and innovation ability are improved through the training and model optimization of large-scale data. At the same time, explainable artificial intelligence solves the "black box" problem and provides designers with a transparent design basis. Secondly, "image classification" (#0) and "scene recognition technology" (#17) play an essential role in the digital protection of cultural heritage furniture. These technologies can efficiently classify traditional furniture styles and optimize the cultural adaptability of the design in combination with historical and cultural scenes. By simulating the human brain's neural structure, the neuromorphic platform promotes the application of real-time processing and intelligent decision-making in furniture design, providing support for design optimization and user demand matching. In addition, "intangible cultural heritage protection and inheritance" (#19) remains a core topic. With the assistance of AI, traditional crafts can be digitally archived and innovatively applied. Finally, statistical analysis tools, such as the "Spearman correlation coefficient" (#3), are essential in understanding the relationship between design variables and optimizing decisions [6], while network security technology provides security for digital cultural heritage protection. The integration and development of these technologies have injected continuous innovation momentum into Traditional cultural heritage furniture design and protection.

From a Chinese perspective, according to keyword clustering Fig. 3b, cross-research on traditional cultural heritage furniture and artificial intelligence covers multiple research topics, reflecting multi-level development from cultural inheritance to technological innovation. The main clusters include "Intangible Cultural Heritage" (#0), "Furniture Design" (#1), "Traditional Craftsmanship" (#11), etc., focusing on the integration of traditional culture and modern design, emphasizing the integration of innovative design and new technologies. Material technology improves the ability to inherit and market the application of cultural heritage. At the same time, keywords such as "design innovation" (#4), "industrial design" (#6) and "panel-type custom furniture" (#7) reflect the trend of modular and customized design, pushing traditional furniture towards functionality and functionality—development with equal emphasis on flexibility. In addition, related research on "Kansei engineering" (#10), "Kansei imagery" (#14) and "grounded theory" (#17) focuses on the analysis of user experience and emotional needs, optimizing designs to meet user preferences better. At the technical level, the combination of "mortise and tenon structure" (#8), "parametric" (#9) and "artificial intelligence" (#12) has significantly improved the structural optimization and production efficiency of traditional furniture. In contrast, "augmented reality" (AR) (#13) and "virtual reality" (VR) (#15) technologies improve market acceptance through virtual environment-assisted design optimization and user interaction. Finally, research related to "Volatile Organic Compounds" (VOC) (#16) focuses on green design and environmentally friendly technology, promoting the sustainable development of the furniture industry through environmentally friendly materials and process optimization. These research topics jointly reflect the comprehensive exploration and development direction of cultural protection, user experience, technological innovation and green design in traditional cultural heritage furniture.

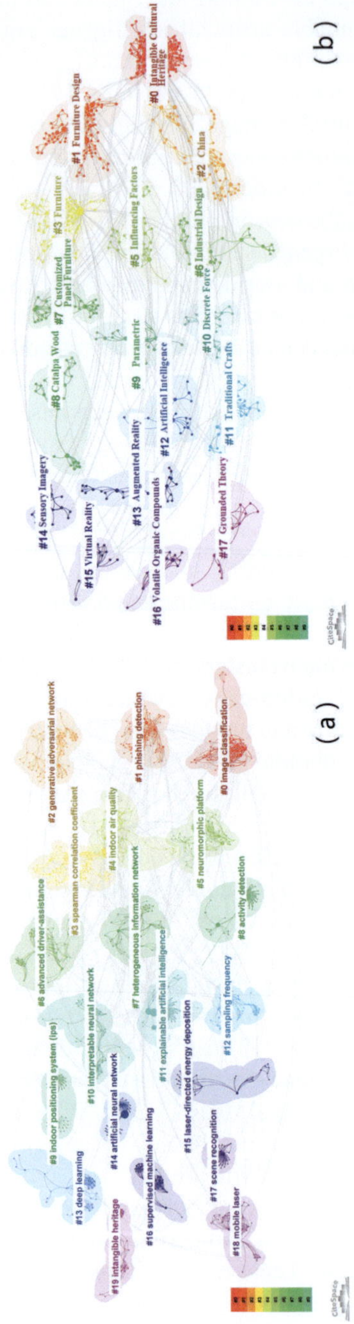


**Fig. 3** **a** Keyword clustering diagram of cross research between global traditional cultural heritage furniture and artificial intelligence. **b** Keyword clustering diagram of cross research between traditional cultural heritage furniture and artificial intelligence

### 3.4 Evolution of Hot Topics and Cutting-Edge Insights

**Evolution of Hot Topics.** The Sankey diagrams (Fig. 4a)of the evolution trend of global research fields show that between 2010 and 2019, the research topics of the international academic community mainly focused on two core directions: cultural heritage protection and intelligent technology application. During this period, artificial intelligence technologies such as deep learning and convolutional neural networks (CNN) rose rapidly and became the main driving force for the intelligent design of traditional cultural heritage furniture. These technologies are widely used in image recognition and design optimization of traditional cultural heritage furniture. At the same time, applying finite element analysis in material optimization and furniture structure design helps optimize furniture's physical strength and service life through simulation technology. The characteristics of international research during this period are mainly technology-driven, and the innovation and optimization of traditional processes are achieved through advanced computing technology and algorithm models.

From 2020 to 2024, global research hotspots have changed significantly. Keyword emergence analysis reveals that Generative Adversarial Networks (GAN) have become a hot technology at this stage, especially in virtual design generation and automated process optimization, which have made breakthrough progress. Applying GAN technology can automatically generate furniture styles and promote the development of personalized customization and design innovation, further expanding the innovation boundaries of cultural heritage furniture design. In addition, the rise of intelligent technologies such as the Internet of Things (IoT) and Digital Twin has promoted the integration of furniture design and intelligent environment interaction. Through the Internet of Things, furniture can realize environmental perception and status monitoring, improving the intelligence level of furniture, which provides new development space for the modern innovation of traditional cultural heritage furniture. At the same time, sustainable development has gradually become the focus of global research after 2020. Researchers increasingly focus on reducing resource consumption and environmental burden through design optimization and material innovation.

As illustrated in Fig. 4b, between 2010 and 2019, Chinese research concentrated on two principal themes: the safeguarding of traditional cultural heritage and the revitalization of traditional crafts. The use of keywords such as "traditional crafts", "cultural heritage", and "furniture design" demonstrates the sustained interest of the academic community in this field. During this period, researchers paid more attention to how to inherit and revive in cultural heritage crafts through modern technology, especially to achieve innovative development driven by digital and intelligent technologies. The introduction of computer-aided design (CAD) enables researchers to integrate modern aesthetics with traditional crafts in the design process, thereby optimizing product design while retaining cultural connotations. In addition, the application of CAD technology has improved design efficiency and promoted the modernization of traditional crafts.

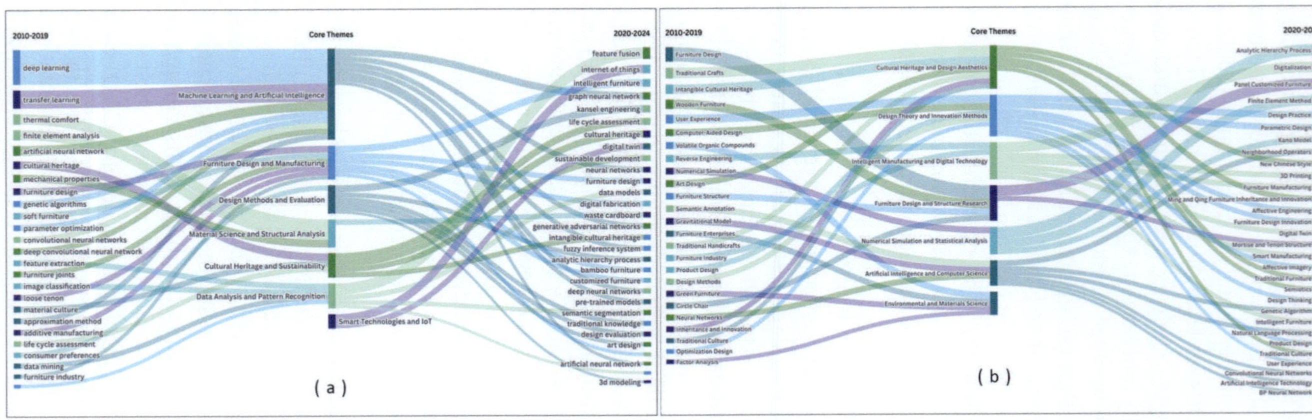


**Fig. 4** **a** Sankey diagram illustrating global research hotspots and trends from 2010 to 2024. **b** Sankey diagram illustrating research hotspots and trends in China from 2010 to 2024

Since 2020, China's research focus has gradually shifted to intelligent manufacturing and design innovation. Emerging words such as Kano model, finite element analysis and Kansei engineering reflect the academic community's increasing attention to the application of intelligent technology in furniture customization and user experience optimization. In particular, applying the Kano model in user demand analysis helps researchers effectively identify and meet users' personalized needs in customized furniture, thereby improving the market acceptance of the design. At the same time, the introduction of Kansei engineering marks the further deepening of design research from the simple appearance and function level to the field of emotional design [7]. In terms of sustainable development, Chinese researchers have gradually turned their attention to the field of green design and environmentally friendly manufacturing. In line with international trends, the research focuses on achieving the goals of resource conservation and environmental protection through using new materials and improving manufacturing processes.

**Analysis of Evolutionary Characteristics**. The evolution of global research in this field presents seven core themes (see Fig. 5a), including cultural heritage and sustainability, data analysis and pattern recognition, furniture design and manufacturing, material science and structural Analysis, artificial intelligence and machine learning, design methods and evaluation, and intelligent technology and the Internet of Things. The evolution and expansion of these sectors reflect the current research hotspots and point out the direction for future research, especially in intelligent design, personalized customization, and green manufacturing, where more breakthrough results are expected to be produced.

The evolution trend of domestic research in China shows six core themes that are slightly different from international research (see Fig. 5b), including cultural heritage and design aesthetics, intelligent manufacturing and digital technology, design theory and innovation methods, furniture design and structural research, numerical simulation and statistical analysis, and artificial intelligence and computers. These research sectors reflect the multi-dimensional exploration of the innovative design of cultural heritage furniture by the Chinese academic community and also point out new development directions for future research.

In general, through the comparative analysis of the research topics of innovative design of traditional cultural heritage furniture in the international and domestic fields, it can be seen that although the two have many similarities in the fields of technology application and design innovation, there are specific differences in the research focus and technology implementation path.

The primary trend of international research focuses on using cutting-edge intelligent technologies to promote automation, personalized customization, and an intelligent interactive environment in furniture design. Technologies such as generative adversarial networks (GAN), the Internet of Things (IoT), and digital twins are widely used in international research and have gradually become the core of intelligent design and manufacturing processes. Global research pays more attention to achieving complex creative generation through deep learning and generative models

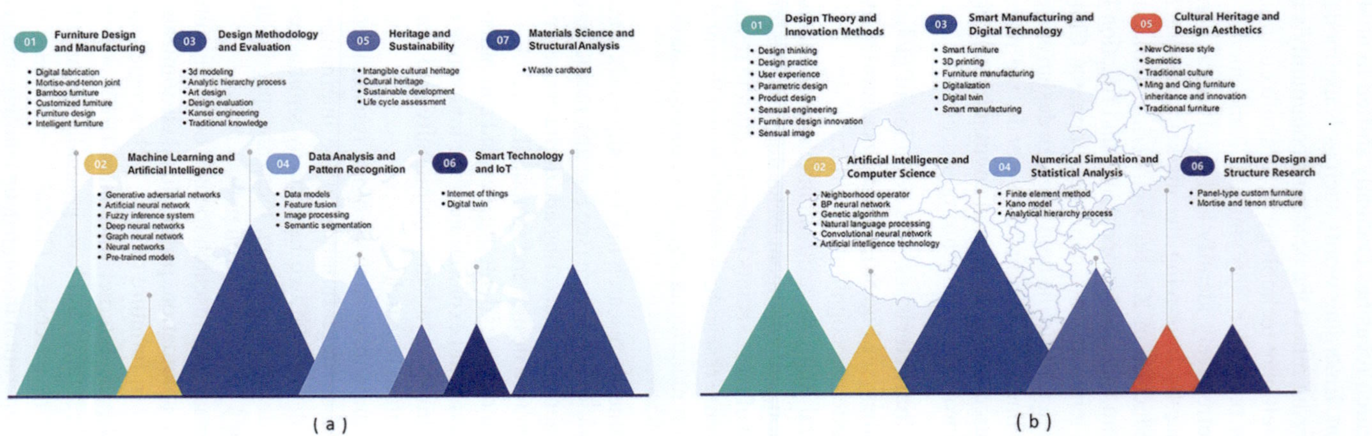


**Fig. 5** **a** Research trends of seven core themes in the innovation field of global traditional cultural heritage furniture. **b** Trend chart of six core themes in the innovation field of traditional cultural heritage furniture in China

in furniture design, thereby accelerating the automation and intelligence of furniture design.

In contrast, Chinese research emphasizes achieving the organic combination of cultural inheritance and innovation in the context of intelligent manufacturing. Chinese research focuses on enabling the revival of traditional craftsmanship and cultural connotations through modern technology. At the same time, Chinese research emphasizes using cultural symbols to protect cultural heritage and innovative design. Many studies focus on integrating traditional symbols, patterns, and structures into modern furniture design to maintain the cultural characteristics of traditional cultural heritage furniture.

# 4 Core Topics and Technical Insights

Based on the previous bibliometric analysis results, this chapter focuses on the research on generative models and innovative design in cultural heritage furniture design and conducts an in-depth analysis of the selected core literature. Based on the in-depth reading and analysis of 103 selected literature, this study summarizes the application of generative models in the field of traditional cultural heritage furniture design into three core aspects: the interaction between generative models and cultural heritage, the application of generative models and innovative design, and the integration of generative models and furniture design.

Various generative models generate new samples by learning data distribution characteristics and have been widely used in various design scenarios [8]. The specific application of generative models mainly revolves around the following models: generative adversarial networks (GANs), variational autoencoders (VAE), normalizing flow, diffusion models, and autoregressive models.

## *4.1 Interaction Between Generative Models and Cultural Heritage*

The application of generative models in the cultural heritage field is gradually deepening, especially with the rapid development of deep learning technologies such as generative adversarial networks (GANs) [9], which have provided new technical means for the digital protection and reproduction of cultural heritage. In recent years, researchers have explored how to use the potential of GANs and their variant models in cultural revival, restoration, re-creation, and virtual display.

The application of generative models has achieved fruitful results in digital restoration. Cortana et al. [10] proposed a GAN-based Dunhuang mural image restoration algorithm, which performed well in repairing damaged mural images, demonstrating the potential of generative models in repairing complex image content. Wang et al.

[11] made progress in the three-dimensional display of ancient cultural relics, using GAN to generate high-resolution three-dimensional traditional cultural heritage models, improving the presentation of cultural relics in virtual environments.

Generative models also provide broad possibilities for cultural re-creation and innovative expression. For example, Shen et al. [12] developed a conditional generative model (TC-StyleGAN2) for multiple target domains based on StyleGAN2, which achieved the generation and creative design of Peking Opera masks, enabling traditional cultural symbols to be presented in a new way. Yang et al. [13] developed a shadow puppet image generation system, SmoothGAN, based on human–computer co-creation, which was used for the generation of Chinese shadow puppet art, supporting the personalized creation needs of users and providing an innovative path for the modern expression of cultural heritage.

In addition to digital restoration and innovative expression, generative models also show unique advantages in the virtual reconstruction and digital protection of cultural heritage. Lu et al. [14] studied the cultural and creative design path of Dunhuang murals and used deep learning technology to achieve digital reproduction in the creative design of murals; Kumar et al. [15] proposed an improved GAN architecture for the restoration of damaged artworks, especially in dealing with artworks damaged by natural and human factors. Hu et al. [16] proposed ConvSRGAN for super-resolution restoration of traditional Chinese landscape paintings, which can perform high-precision magnification and detail restoration of landscape paintings, providing a new method for high-definition display of artworks.

From the perspective of technological evolution, the application of generative models in cultural heritage has gradually shifted from essential image generation and restoration to multi-task, multi-modal generative architectures, forming a higher level of intelligent application. For example, Chung et al. [17] boundary-enhanced generative adversarial network (BEGAN) combines cycle-consistent GAN and pix2pix networks to solve the boundary blur problem in the generation process of traditional landscape paintings, significantly improving the clarity of the generated effect. The calligraphy character generation model combining dense blocks and capsule networks developed by Zhang et al. [18] can generate calligraphy characters that meet aesthetic requirements in an end-to-end mode. These advances reflect the increasing complexity and application depth of generative models and the increasing diversification of model design, which meets the protection and re-creation needs of different cultural heritages.

From the perspective of cultural significance, generative models have injected innovative vitality into cultural heritage. In protecting red revolutionary heritage, Yi et al. [19] developed a GAN-based image restoration model that can restore images of revolutionary heritage damaged by corrosion and oxidation, which has significant historical and cultural value. In terms of the re-creation of traditional crafts, Wang et al. [20] proposed a creative design model for rattan patterns based on a deep convolutional generative adversarial network (DCGAN), which provides technical support for the combination of traditional handicrafts and modern design.

However, the application of generative models in cultural heritage also faces some challenges, including the accurate expression of complex cultural symbols and craft details by the model and maintaining cultural authenticity and aesthetic consistency during the generation process. Goodarzi et al. [21] pointed out that in the case of scarce data, generative models have particular difficulties in accurately reproducing early design decisions and cultural characteristics. Future research can focus on the structural optimization of generative models, the potential of new techniques like cultural diffusion models, and the improvement of the recovery and generation functions of the models. Sukkar et al. [22] conducted a preliminary evaluation of the application effect of diffusion models in the virtual reconstruction of Islamic architectural heritage, showing the broad prospects of this model in cultural revitalization.

In summary, the application of generative models in cultural heritage has expanded from single-image generation to multimodal complex generation tasks. With the development of technology, generative models will continue to promote the digital revival of cultural heritage and bring new opportunities and challenges to the cultural and creative industries. In future research, exploring the innovative application of multimodal fusion, intelligent model optimization, and diffusion models of generative models will become an essential direction for further promoting the digitization and modernization of cultural heritage.

## *4.2 Application of Generative Models and Innovative Design*

In the cross-application field of generative models and design innovation, intelligent generative models have become a necessary technical means to promote innovation in design sub-fields. These technologies have shown broad application potential in visual communication, industrial design, fashion design, architectural design, and other fields, and they have gradually formed an interdisciplinary, innovative theoretical framework for design.

Generative adversarial networks (GANs), as a generative model, are widely used in art creation, product design and engineering design due to their powerful generative capabilities. For example, The Seg2pix model proposed by Seo et al. [23] is based on GAN technology and provides accurate image conversion capabilities for animation design. In addition, Wang et al. [24] applied GAN in architectural structure design and proposed a new method that can help design decisions by combining parameter optimization of manufacturing process constraints.

In recent years, diffusion models have gradually emerged in design generation, providing generation capabilities and progressive generation process advantages different from GAN. The diffusion model proposed by Chen et al. [25] generates high-quality architectural exterior designs through text prompts, demonstrating the creative application of AI in architectural design. The RealtimeGen tool developed by Li et al. [26], combined with Adobe Photoshop, supports the real-time generation of commercial digital art, helping designers to make dynamic adjustments during the creative process. Maksoud et al. [27] used the AI image generation tool Midjourney to design a mosque based on the Safavid architectural style, exploring the potential of AI in religious building design. Diffusion models have unique advantages in specific fields such as architecture, art, and product design. They can generate high-resolution images with rich details, providing a new solution for complex design generation.

With the development of generative models, new application trends are also emerging. For example, Lee et al. [28] used GPT-4 and DALL-E-3 to generate design sketches in eco-design automatically. Through a tool called Eco-innovate Assistant, text solutions and sketches were generated based on the design problem statement provided by the user, demonstrating the application prospects of large language models (LLMs) in eco-design.

Whether art creation, fashion, architecture, or automobile design, various generative models demonstrate different application scenarios and innovative results. Each research case demonstrates the flexibility and diversity of generative models and shows their potential to improve design efficiency, enrich design expression, and support creative generation. These studies provide practical technical paths for the development of the future design industry and reveal the broad applicability of generative models in different design fields. Analysing generative model application examples shows how intelligent generation technology can achieve interdisciplinary integration through innovative applications, open up new prospects for design innovation and automation, and provide rich references for academic research and practical applications.

## 4.3 Integration of Generative Models and Furniture Design

In furniture design, applying generative models has become an essential technical force to promote design innovation and improve design efficiency. With the help of cutting-edge technologies such as deep learning, generative adversarial networks (GAN), diffusion models, and optimization algorithms, furniture design has made significant progress in automation, personalization, and sustainability.

Generative adversarial networks (GAN) are widely used in furniture design and have gradually become a powerful tool for designers to carry out conceptual and generative designs. The SDM-NET model proposed by Gao et al. [29] is based on variational autoencoders (VAE) and is used to generate three-dimensional models of product designs. Extracting the latent space features of design samples achieves the automatic generation of different design styles. And iterate quickly. The stylized scene-enhanced GAN (SSE-GAN) proposed by Zhang et al. [30] can generate diversified interior design solutions and enhance specific styles, making it capable of adapting to different decoration styles. Applying such models in interior design and furniture layout demonstrates the unique advantages of generative models in the diversity and personalization of design scenarios.

Generative models are also crucial in optimizing residential floor plans and spatial layouts. As an illustration, Jahan et al. [31] proposed a semantic-guided latent space exploration method (Semantics-Guided Latent Space Exploration), which allows designers to explore the latent space of furniture at the semantic level, thereby achieving more flexible furniture. Design. Similar methods have also been applied in the field of industrial design. For example, the Info-Giraffe model proposed by Zhou et al. [32] uses GAN to generate designs with virtual neural features, allowing furniture's functional and aesthetic design to be better integrated.

Generative models show application potential in emerging furniture design and lead new design methods in the innovative design of traditional furniture. Kang et al. [33] proposed a framework combining deep convolutional neural network (DCNN) and DCGAN for the design generation of bamboo furniture. This framework leverages the generative power of GAN and designers' creativity through the collaboration of AI and designers to give traditional bamboo furniture greater flexibility in design style and form. Wang Z et al. [34] further proposed the Interact 2.0 system, which integrates constructed solid geometry (CSG) and deep learning (DL) to generate complex furniture geometric models and achieve a high degree of automation in furniture generation. This system demonstrates the advantages of generative models in morphology generation and solid modelling in innovative furniture design and promotes the development of automated furniture design.

Diffusion models, an emerging technology in generative models, are gradually emerging in furniture design. Xu et al. [35] proposed the BrepGen model, which generates three-dimensional furniture designs based on diffusion models. The progressive generation process allows furniture shapes and complex structures to be generated with great detail, expanding furniture design possibilities. Wang Y. et al. [36] developed the MFMP toolkit based on generative AI for innovative design and sustainable exploration of traditional furniture. The toolkit combines the progressive generation capability of the diffusion model, enabling designers to make modern innovations based on traditional furniture while ensuring the environmental protection and sustainability of the design, providing a new solution for the innovation and inheritance of traditional furniture design (see Fig. 6).

The application of generative models in furniture design not only stays in the generation stage but also involves optimizing the design process and exploring human–computer collaborative design. Tanasra et al. [37] combined generative models such as pix2pix, BicycleGAN, and SPADE to develop a machine learning-based architectural space optimization method that helps optimize space configuration in furniture design. By automatically generating a variety of spatial layouts and

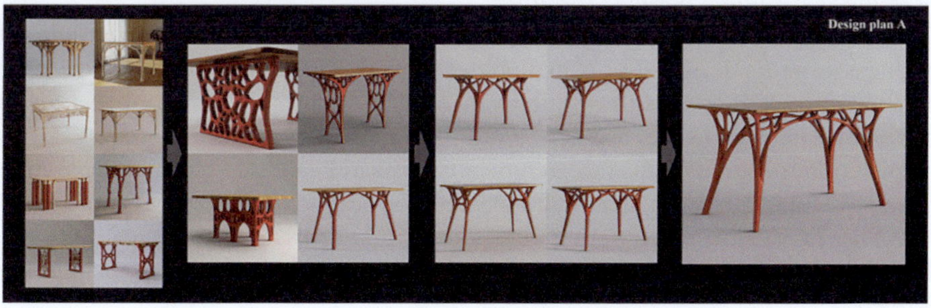


**Fig. 6** Ming-style furniture generative innovative design: Based on the MFMP toolkit [36]

furniture configuration solutions, the generative model significantly improved the efficiency and intelligence of space design.

Generative models have demonstrated diverse application scenarios and innovative results in sub-fields such as product design, interior design, furniture model generation, and traditional furniture innovation. These research examples demonstrate that generative models have brought unprecedented automation and intelligent development to furniture design, laying a solid foundation and supporting future design innovation. Through in-depth analysis of these cases, we can see how generative models play a core role in process optimization, design innovation, and human–machine collaboration in furniture design, providing a solid technical foundation and broad application prospects for the further development of the design industry.

## 5 Results and Discussion

Generative artificial intelligence technology has achieved remarkable results in cultural heritage and innovative design. Researchers use algorithms to create new design solutions and effectively inherit cultural characteristics. However, as a field closely related to people's daily lives, furniture design lags in the application research of generative artificial intelligence technology. Research in this field is particularly scarce, especially in the innovative design of traditional cultural heritage furniture, because it involves a deep understanding of traditional culture and the inheritance of unique craftsmanship. Therefore, to better promote the development of furniture design, primarily to protect and inherit the precious traditional heritage furniture culture, it is necessary to strengthen the research on generative artificial intelligence in furniture design, especially in the innovative design of cultural heritage furniture. More innovative design methods and approaches can be explored by deeply exploring and analyzing cultural heritage furniture's unique elements and craftsmanship characteristics, combined with the advantages of generative artificial intelligence technology. It will not only help to enhance the innovation and artistic value of furniture design but also effectively promote the inheritance and development of traditional culture and inject new vitality and possibilities into the innovative design of traditional cultural heritage furniture.

This study conducted a systematic quantitative analysis of the literature on cultural heritage furniture design, deeply explored the application scenarios and prospects of generative models and artificial intelligence in the innovative design of traditional cultural heritage furniture, and sorted out the core literature and technology development trends. Based on the research results, a theoretical research path map for the generative innovative design of traditional cultural heritage furniture was constructed (see Fig. 7). This research path includes three main parts: existing achievements, core technologies, and future research trends, and the following key conclusions are formed:

**Existing research results: Integration of technology and culture in multi-dimensional fields**. Current research on traditional cultural heritage furniture

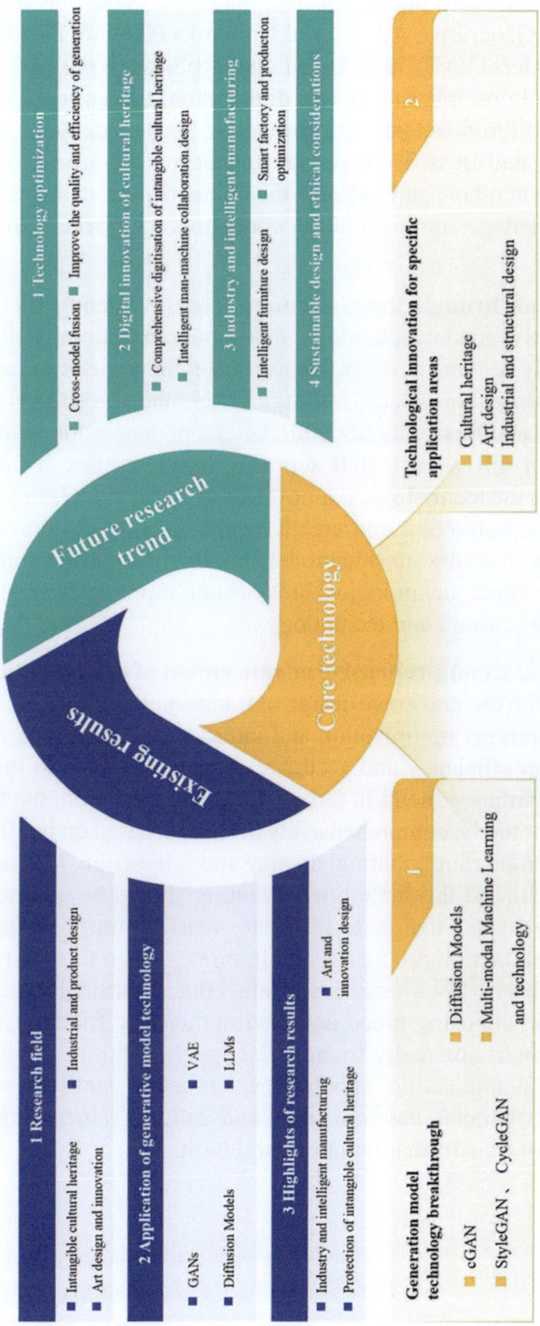


**Fig. 7** Theoretical research path chart of the generative innovative design of traditional cultural heritage furniture (Drawn by the author)

involves three main areas: traditional cultural heritage protection, art and design innovation, and industrial and intelligent manufacturing. With the support of core technologies such as Generative Adversarial Networks (GANs), Diffusion Models, Variational Autoencoders (VAE), and Large Language Models (LLMs), research has been carried out in cultural inheritance and digitization, design method innovation, and industrialization. Significant advancements have been made in intelligent manufacturing and other related areas. These research results promote the digital protection of traditional cultural symbols and enhance the competitiveness and adaptability of traditional cultural heritage furniture in the contemporary market through modern technological means.

**Core technology breakthrough: innovative application of generative models**. The application of generative model technology in designing traditional cultural heritage furniture has gradually become a research hotspot. Technologies such as conditional generative adversarial networks (cGAN), StyleGAN, and CycleGAN significantly improve the accuracy and flexibility of cultural element generation, while the multimodal combination of diffusion models supports more complex design tasks. The research also explores the technological innovation of these models in specific fields, including the accurate extraction and creative generation of cultural heritage, the expansion of diversity in art design, and model data-driven optimization of industrial and structural design. These technological breakthroughs provide reliable support for the deep integration of culture and technology.

**Future research trends: comprehensive advancement of digitalization and intelligence**. Future research on innovative design of traditional furniture cultural heritage should focus on technology optimization and interdisciplinary cooperation. On the technical level, design efficiency and accuracy should be improved through model fusion and algorithm improvement; in terms of digital innovation, the digitalization of cultural heritage should be comprehensively promoted based on intelligent design, from data collection, modeling to virtual display and interaction, to form a complete digital ecological chain. At the same time, attention should be paid to the realization of dynamic inheritance, that is, to integrate smart furniture design with smart factories, optimize production processes, and improve design flexibility and market adaptability. On this basis, sustainable design and ethical considerations should also be emphasized: while choosing green and environmentally friendly materials and paying attention to environmentally friendly design, the cultural values and ethical issues in technological applications should be examined in depth to ensure a multidimensional balance of social, environmental and cultural values in the process of promoting the modern transformation of cultural heritage.

## References


1. Ning, G.: The inheritance and innovation of Chinese creation culture. Res. Literary Art Theor. **39**(02), 1–12 (2019)
2. Dongning, B., Guangqun, Y., Ming, C.: Research and practice on the protection and inheritance of the intangible cultural heritage of "Ming-style furniture making skills"—the basic construction of the library of miniature models of ancient Chinese furniture in museums. J. Shandong Inst. Arts Crafts **06**, 106–110 (2018)
3. Xiaona, M., Tula, X.Y.: Current status of digital development of intangible cultural heritage. Chinese Sci. Inf. Sci. **49**(02), 121–142 (2019).
4. Chen, C., Song, M.: Visualizing a field of research: a methodology of systematic scientometric reviews. PLoS ONE **14**(10), e0223994 (2019)
5. Liang, Z., Xianxue, M.: Comparative study of bibliometrics and content analysis. Lib. Work Res. **06**, 64–66 (2013)
6. Xue, G., Chen, J.: Strategies for applying shape grammar to wooden furniture design: taking traditional Chinese Ming-style recessed-leg table as an example. BioResources, **19**(1) (2024)
7. Lin, Q., Cai, J., Xue, Y.: Affective response difference to the viewing of different styles of solid wood furniture based on kansei engineering. BioResources, **19**(1) (2024)
8. Morris, M.R., Cai, C.J., Holbrook, J., et al.: The design space of generative models. arXiv preprint arXiv:2304.10547 (2023)
9. Goodfellow, I., Pouget-Abadie, J., Mirza, M., et al.: Generative adversarial networks. Commun. ACM **63**(11), 139–144 (2020)
10. Ciortan, I.M., George, S., Hardeberg, J.Y.: Colour-balanced edge-guided digital inpainting: applications on artworks. Sensors **21**(6), 2091 (2021)
11. Wang, J., Song, J., Zhang, Y., et al.: Design of 3D display system for intangible cultural heritage based on generative adversarial network. Sci. Program. **2022**(1), 2944750 (2022)
12. Shen, Y., Duan, O., Xin, X., et al.: Styled and characteristic Peking opera facial makeup synthesis with co-training and transfer conditional styleGAN2. Heritage Sci. **12**(1), 358 (2024)
13. Yang, C., Lou, M., Chen, X., et al.: An interactive collaborative creation system for shadow puppets based on smooth generative adversarial networks. Comput. Mater. Continua, **79**(3) (2024)
14. Lu, X., Yang, Y., Wang, L., et al.: Cultural and creative design of Dunhuang murals in Xixia Period based on deep learning. Wirel. Commun. Mob. Comput. **2022**(1), 7161593 (2022)
15. Kumar, P., Gupta, V., Grover, M.: Dual attention and channel transformer based generative adversarial network for restoration of the damaged artwork. Eng. Appl. Artif. Intell. **128**, 107457 (2024)
16. Hu, Q., Peng, X., Li, T., et al.: ConvSRGAN: super-resolution inpainting of traditional Chinese paintings. Heritage Sci. **12**(1), 176 (2024)
17. Chung, C.Y., Huang, S.H.: Interactively transforming Chinese ink paintings into realistic images using a border enhance generative adversarial network. Multimedia Tools Appl. **82**(8), 11663–11696 (2023)
18. Zhang, W., Sun, Z., Wu, X.: An end-to-end generation model for Chinese calligraphy characters based on dense blocks and capsule network. Electronics **13**(15), 2983 (2024)
19. Yi, J., Tian, Y., Zhao, Y.: Novel approach to protect red revolutionary heritage based on artificial intelligence algorithm and image-processing technology. Buildings **14**(9), 3011 (2024)
20. Wang, T., Ma, Z., Yang, L.: Creativity and sustainable design of wickerwork handicraft patterns based on artificial intelligence. Sustainability **15**(2), 1574 (2023)
21. Goodarzi, P., Ansari, M., Rahimian, F.P., et al.: Incorporating sparse model machine learning in designing cultural heritage landscapes. Autom. Constr. **155**, 105058 (2023)
22. Sukkar, A.W., Fareed, M.W., Yahia, M.W., et al.: Analytical evaluation of Midjourney architectural virtual lab: defining major current limits in AI-generated representations of Islamic architectural heritage. Buildings **14**(3), 786 (2024)
23. Seo, C.W., Seo, Y.: Seg2pix: few shot training line art colorization with segmented image data. Appl. Sci. **11**(4), 1464 (2021)

24. Wang, Y., Du, W., Wang, H., et al.: Intelligent generation method of innovative structures based on topology optimization and deep learning. Materials **14**(24), 7680 (2021)
25. Chen, J., Wang, D., Shao, Z., et al.: Using artificial intelligence to generate master-quality architectural designs from text descriptions. Buildings **13**(9), 2285 (2023)
26. Li Z, Zhang Y, Zhou S et al (2024) RealtimeGen: an intervenable ai image generation system for commercial digital art asset creators. Int. J. Human–Comput. Interact. 1–24 (2024)
27. Maksoud, A., Elshabshiri, A., Saeed Hilal Humaid Alzaabi, A., et al.: Integrating an image-generative tool on creative design brainstorming process of a safavid mosque architecture conceptual form. Buildings, **14**(3), 843 (2024)
28. Lee, C.K.M., Liang, J., Yung, K.L., et al.: Generating TRIZ-inspired guidelines for eco-design using generative artificial intelligence. Adv. Eng. Inform. **62**, 102846 (2024)
29. Gao, L., Yang, J., Wu, T., et al.: SDM-NET: Deep generative network for structured deformable mesh. ACM Trans. Graph. (TOG) **38**(6), 1–15 (2019)
30. Zhang, S., Han, Z., Lai, Y.K., et al.: Stylistic scene enhancement GAN: mixed stylistic enhancement generation for 3D indoor scenes. Vis. Comput. **35**, 1157–1169 (2019)
31. Jahan, T., Guan, Y., Van Kaick, O.: Semantics-guided latent space exploration for shape generation. Comput. Graph. Forum. **40**(2), 115–126 (2021)
32. Zhou, F., Zhang, J., Jiang, Y., et al.: Info-Giraffe: virtual neural feature field generation for object in limited views condition. J. Electron. Imaging **32**(2), 023009–023009 (2023)
33. Kang, X., Nagasawa, S., Wu, Y., et al.: Emotional design of bamboo chair based on deep convolution neural network and deep convolution generative adversarial network. J. Intell. Fuzzy Syst. **44**(2), 1977–1989 (2023)
34. Wang, Z., Deng, Y., Yang, J., et al.: Generative deformable radiance fields for disentangled image synthesis of topology-varying objects. Comput. Graph. Forum. **41**(7), 431–442 (2022)
35. Xu, X., Lambourne, J., Jayaraman, P., et al.: Brepgen: a b-rep generative diffusion model with structured latent geometry. ACM Trans. Graph. (TOG) **43**(4), 1–14 (2024)
36. Wang, Y., Xi, Y., Liu, X., et al.: Exploring the dual potential of artificial intelligence-generated content in the Esthetic reproduction and sustainable innovative design of Ming-style furniture. Sustainability **16**(12), 5173 (2024)
37. Tanasra, H., Rott Shaham, T., Michaeli, T., et al.: Automation in interior space planning: utilizing conditional generative adversarial network models to create furniture layouts. Buildings **13**(7), 1793 (2023)

# Intelligent Image Analysis and Detection Technology

# Surface Defect Detection Model for Quick-Connect Fittings

**Lin Zhou, Chen Wang, Shuai Yang, and Peng Huang**

**Abstract** In response to issues such as missed detection of small defects, low detection efficiency for certain defects, and insufficient overall automation in the automated detection of Quick-Connect Fittings (QCF), this paper proposes an intelligent detection model based on the collaborative use of supervised and unsupervised learning. First, a supervised training approach is conducted using an improved YOLOv8 model (QCF-YOLO), which optimizes the feature map representation capability, significantly improving the detection accuracy of small defects. This model has the advantages of fewer parameters and easy deployment, meeting real-time detection requirements. Secondly, unsupervised secondary training is performed in the cloud using detection results and raw images collected from edge devices. By comparing the detection results of supervised and unsupervised models, the model's ability to detect unknown defect categories is progressively enhanced, effectively avoiding false detections caused by background interference in unsupervised training. Finally, based on a "cloud-edge-end collaborative architecture", the model is deployed at the edge. Through the collaborative work of cloud training, edge detection, and terminal data collection, the automation level and detection efficiency of Quick-Connect Fitting defect detection are significantly improved.

**Keywords** Deep learning · Small sample learning · Small sample target detection · Target detection

L. Zhou · C. Wang (✉) · S. Yang · P. Huang
School of Automotive Intelligent Manufacturing, Hubei University of Automotive Technology, Shiyan, China
e-mail: 20090011@huat.edu.cn

L. Zhou
e-mail: 202311100@huat.edu.cn

S. Yang
e-mail: 793132180@qq.com

P. Huang
e-mail: 2037721553@qq.com



Y. Ma (ed.), *Proceedings of 4th International Conference on Intelligent Systems Design and Engineering Applications*, Lecture Notes in Electrical Engineering 1478,
https://doi.org/10.1007/978-981-95-1802-9_10

## 1 Introduction

Quick-connect fittings (QCF) serve as practical elements within automotive fluid systems and are vital to the functionality of air brake mechanisms. The connections in brake lines require a secure seal, as any leakage can result in brake system failures [1]. Thus, ensuring the airtight integrity of QCFs is of utmost importance. After assembling a quick-connect fitting, it is imperative to conduct thorough defect inspections.

At present, the detection of defects in QCFs primarily depends on manual visual inspections. This process usually begins with a basic airtightness evaluation [2] that determines if any flaws exist that could compromise sealing effectiveness, followed by a visual review of potential defects, including indentations, fractures, misalignment of the central hole, and discoloration. Nevertheless, many defects in quick-connect fittings are minor, categorizing them as small-object detection issues. The manual inspection process is susceptible to inaccuracies due to visual fatigue and generally exhibits low efficiency [3]. With advancements in digital image processing technologies, algorithms utilizing machine vision for defect identification are becoming increasingly prevalent in multiple sectors [4]. For instance, Zhang et al. [5] presented an algorithm based on multi-exposure fusion aimed at highlighting defects in automotive brake hose joints, though it primarily serves as a preprocessing step rather than facilitating direct defect identification. Conventional techniques like threshold segmentation [6] can isolate or emphasize critical information but typically only focus on detecting rudimentary characteristics such as edges, rendering small-object detection difficult [7].

In contrast, deep learning has demonstrated significant effectiveness in object detection and is emerging as a leading technique for intelligent defect identification [8]. Peng et al. [9] introduced a modified Faster-CNN model [10] designed to detect minor defects in polyethylene pipe fusion joints, achieving a detection accuracy of 90.5% while utilizing around 317 million parameters. Likewise, for recognizing defects on sealing surfaces in internal thread joints, Ding et al. proposed an enhanced YOLOv4 detection algorithm, which attained an accuracy rate of 87% with 64 million parameters.

To minimize the necessary storage and computational resources for implementing deep learning models on embedded devices [11], researchers have investigated compressed neural networks. Liao et al. [12] presented a streamlined YOLOX architecture containing 5.03 M parameters to improve the accuracy and efficiency while reducing the computational demands associated with larger models in detecting surface defects. In a related effort, Luo et al. [13] introduced an enhanced YOLOv8 algorithm that integrates Ghost modules and multi-scale attention techniques, achieving a detection accuracy of 93.5% with 5.2 M parameters, successfully implemented on a Raspberry Pi 4B. Furthermore, Zhang et al. [14] devised a collaborative cloud-edge-End model [15] for fabric defect detection with a parameter count of 2.878 M, which was effectively deployed on an NVIDIA Jetson TX2 [16].

Nevertheless, the aforementioned techniques heavily depend on extensive labeled datasets for their training processes. In actual production scenarios, instances of defects are infrequently encountered, while non-defective examples are substantially more prevalent. Unsupervised learning, which involves the analysis of unlabeled data without requiring manual annotations, offers a solution to this challenge [17]. Nejc et al. [18] recommended the Decode AE model, which utilizes incremental learning alongside unsupervised approaches to tackle issues related to dataset labeling and the scarcity of defect images, thereby significantly enhancing the efficiency of industrial visual inspections. Additional studies, including those by Xing Tao et al. [19] and Gyohyeok et al. [20], underscore the effectiveness of applying unsupervised techniques in conjunction with deep learning to improve defect detection and classification outcomes.

In response to the industry's need for an optimal equilibrium among data security, detection efficiency, model accuracy, and automation in QCF defect identification, the following objectives have been established:

(1) Attain an average detection accuracy exceeding 90%.
(2) Enhance the capability to detect minor defects.
(3) Facilitate real-time detection with a frame rate surpassing 30 FPS.
(4) Implement the model on an embedded device (NVIDIA Jetson TX2) for smooth integration into the "cloud-edge-end collaborative architecture".

This paper introduces a model designed to fulfill these criteria, with the aim of advancing the detection of QCF defects while tackling the hurdles related to edge deployment, efficiency, and model accuracy.

## 2 Related Works

### 2.1 Enterprise Testing Process

As shown in Fig. 1, the operational diagram of the current enterprise inspection is presented. In contemporary enterprises, the inspection process for quick plug connectors commences with a fundamental air tightness test. This test begins by connecting the nylon end of the quick plug connector to the ventilation device, ensuring that the threaded pipe connection port is securely sealed with bolts. The product is subsequently immersed in water, and compressed air is injected into the nylon tube. By observing the presence of bubbles around the quick plug joint, products exhibiting partial cracks and eccentric defects in the central hole can be initially identified. Following this, the product is examined for additional defects, including bumps, cracks, eccentricity of the central hole, or white markings.

**Fig. 1** Actual inspection line of the enterprise

## *2.2 Cloud-Edge-End Collaborative Architecture for Inspection of Surface Defects in QCF*

As shown in Fig. 2, the Cloud-Edge-End Collaborative Architecture consists of three key components: the cloud system, the edge and the endpoints. The cloud system, as the core, is responsible for executing complex product model training tasks and collating detection results into a database. The edge is deployed near the data source and carries the cloud-trained models for real-time inference, while performing efficient data collaboration and interaction with the cloud. The end devices are then responsible for collecting data and providing inputs to the whole system. This architecture is designed to optimize the data processing process, reduce latency, improve efficiency, and guarantee immediate data analysis and decision support.

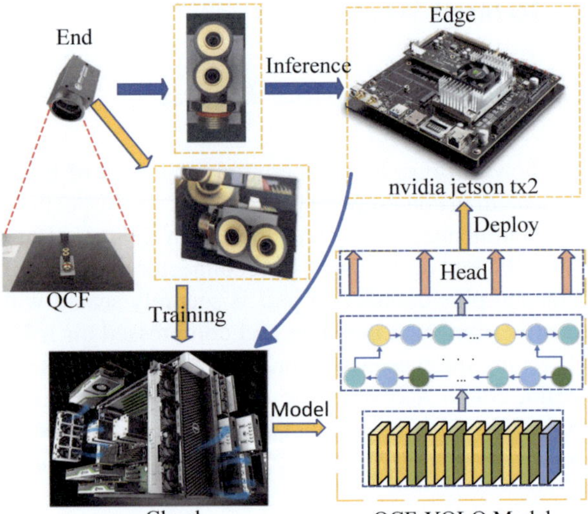


**Fig. 2** Schematic diagram of "cloud-edge-device collaborative architecture" for surface defect detection of QCF

## 2.3 Data

The original data set of this article is provided by a domestic fast plug connector manufacturer and contains 330 defect pictures. There are four typical defects in total, namely white seal, cracking, bumping, and eccentricity of the central hole. Since the mold clamping line is similar to the cracking characteristics, it avoids false detection. This article also regards the mold clamping line as a defect type. The comparison diagram of the mold clamping line and cracking is shown in Fig. 3. Since the amount of defective data in the real scenario is relatively small, considering that the data set will not have strong robustness due to too few samples, the original data set was geometrically transformed. For example, data enhancement methods such as flip, rotate, crop, deformation, and scaling. The number of pictures of the expanded dataset is 2310. The data set is divided into training set, validation set and test set at a ratio of 7:2:1. See Table 1 for specific data.

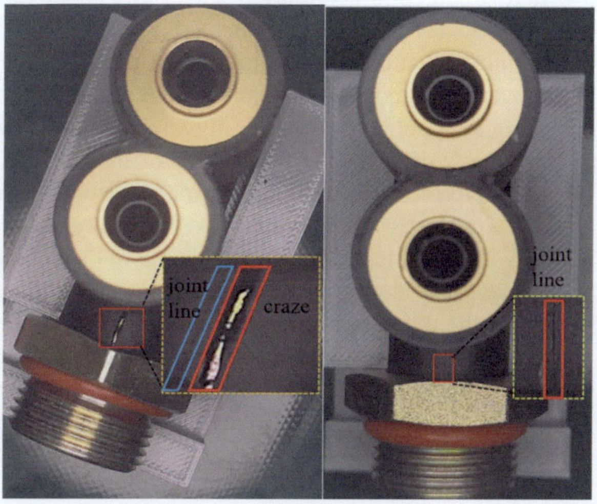


**Fig. 3** Comparison of joint line and craze

**Table 1** Division table for QCF dataset

| Defect type | Train (70%) | Validate (20%) | Test (10%) | Total |
|---|---|---|---|---|
| Bump | 1104 | 307 | 164 | 1575 |
| White mark | 938 | 259 | 126 | 1323 |
| Craze | 731 | 201 | 97 | 1029 |
| Off-centered | 577 | 172 | 91 | 840 |
| Joint line | 1010 | 280 | 138 | 1428 |
| All | 1617 | 462 | 231 | 2310 |

$$
\begin{aligned}
r_f &= \frac{c_1 \times h_1 \times w_1 \times c \times k \times k \times 2}{\frac{c_1}{s} h_1 \times w_1 \times c \times k \times k \times 2 + \frac{c_1}{s} h_1 \times w_1 \times (s-1) \times d \times d \times 2} \\
&\approx \frac{s \times c}{c + s - 1} \approx s
\end{aligned} \quad (1)
$$

$$
\begin{aligned}
r_p &= \frac{c_1 \times (c \times k \times k + 1)}{\frac{c_1}{s} \times (c \times k \times k + 1) + \frac{c_1}{s} \times ((s-1) \times d \times d + 1)} \\
&\approx \frac{s}{1 + \frac{s \times c}{c \times k \times k + 1}} \approx s.
\end{aligned} \quad (2)
$$

# 3 Modeling

In response to the demand for model deployment at the edge and the challenge of low defect detection accuracy in QCF, this paper introduces a lightweight model, QCF-YOLO, for the intelligent detection of QCF defects. The model incorporates a lightweight Ghost Conv convolution module and a C3Ghost module, which replace the conventional convolution and C2f module in the backbone network, thereby achieving a reduction in the overall weight of the network. Additionally, the feature pyramid network (FPN) is optimized for lightweight performance by utilizing the DySample to recover detailed information from low-resolution feature maps. This approach mitigates the accuracy loss typically associated with a reduction in the number of parameters. To enhance the model's capacity for detecting small targets, a small target detection head, P2, is integrated into the head section. This addition facilitates the extraction of feature information from numerous small targets, ultimately improving the model's accuracy in identifying tiny defects. Figure 4 illustrates the structure of the QCF-YOLO model, with the notation '1/2' in the neck section indicating that the number of channels is halved.

## *3.1 C3GNet*

The Ghost Convolution module was proposed by Han et al. [21] in 2020, mainly to address the deployment of neural network models on embedded devices with limited memory and computational resources. The model can generate a larger number of feature maps with a relatively small number of parameters, thereby obtaining the corresponding redundant feature maps at a lower computational cost. The information contained in these redundant feature maps is essential for achieving a comprehensive understanding of the input data. With a given number of input feature maps, Ghost Conv can produce an equivalent number of feature maps as the standard convolution

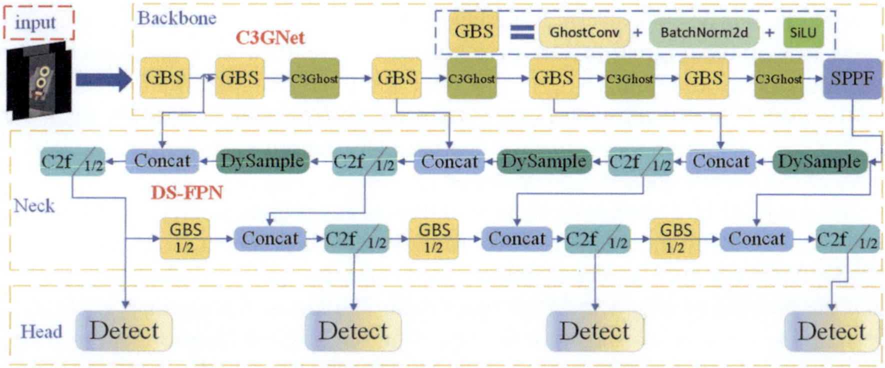


**Fig. 4** Structure diagram of QCF-YOLO

operation through a two-step process. As illustrated in Fig. 5, (a) depicts the ordinary convolutional structure, while (b) shows the Ghost Conv structure.

As illustrated in Fig. 5, the Ghost Conv module comprises three components: Conv, Ghost, and Intrinsic. The Conv module executes standard convolution operations on the input data, resulting in feature map 1. The Ghost feature map is produced

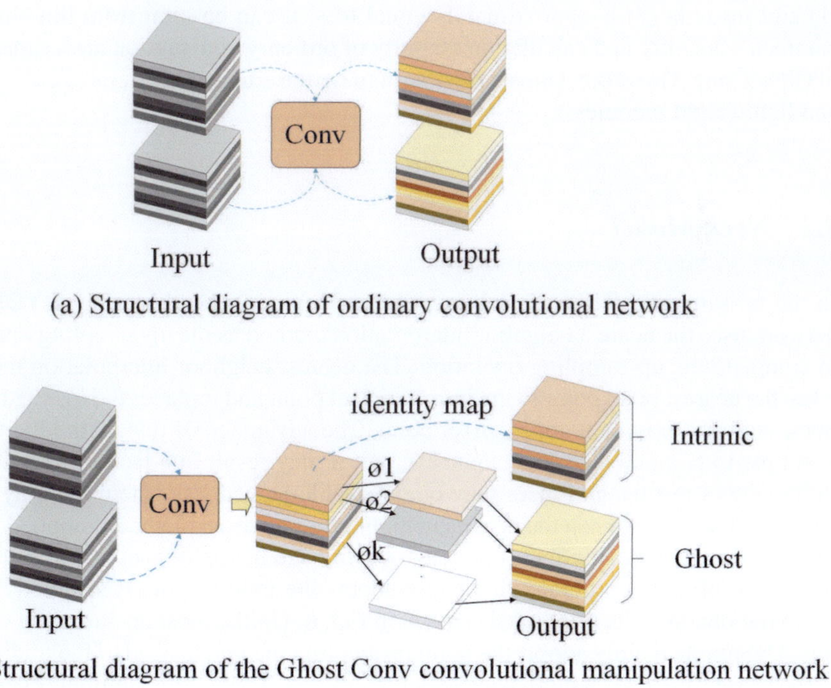


**Fig. 5** Comparison of Ghost Conv and normal convolutional operation network structure

by applying a series of convolutional layers with a kernel size of Φk (typically 3 × 3 or 5 × 5) to the original feature map, capitalizing on their high linear correlation as redundant information. To derive the final output feature map, we first transform feature map 1 into its intrinsic feature representation through an identity map, and subsequently combine these intrinsic feature maps with the Ghost feature maps to enhance the overall feature representation.

As shown in Fig. 5, the Ghost Conv module consists of three parts: Conv, Ghost and Intrinsic. The Conv module performs ordinary convolution operations on the input data to obtain feature Fig. 1. The output channel number of this module requires less than the number of channels of the input data. The Ghost feature map is obtained by applying a series of convolutional layers on the original feature map, in which each layer uses a convolution kernel of size Φk, which is usually 3 × 3 or 5 × 5. Because there is a high degree of linear correlation between these Ghost feature maps and the original feature map, they can be regarded as redundant information of the original feature map. To obtain the final output feature map, we first convert feature map 1 into its eigenfeature representation through an identity map, and then splice these eigenfeature maps with the Ghost feature map, thereby enhancing the feature expression ability. Formula (1) is the ratio of FLOPs of ordinary convolution operations to FLOPs of Ghost module; Formula (2) is the ratio of Params of ordinary convolution module to Ghost module. Since s is the number of transformations, which is totally smaller than the number of input channels, the final result in formula (1) and formula (2) is approximately equal to s. It can be seen from this that the parameter quantity and calculation quantity of ordinary convolution are s times that of GhostConv. Therefore, GhostConv can achieve the effect of accelerated computing and lightweight modules.

## 3.2 Neck Model

In the up-sampling (UP Sample) part of the neck network, the original YOLOv8 network uses the nearest neighbor interpolation method as the up-sampling operator to complete the up-sampling operation. The nearest neighbor interpolation method takes the nearest pixel point to the inserted pixel point and replaces the inserted pixel point with the gray value of its pixel point. The advantage of this method is that it is simple to compute and less expensive, but it may result in a lack of smoothness in the output results, and there may be noticeable jaggedness where the gray level changes. Therefore, when the nearest neighbor interpolation method is applied to the up-sampling process, it affects the image quality and has an impact on the detection results to solve this problem, this paper adopts the lightweight DySample [22] up sampling operator. Its principle is shown in Fig. 6. Unlike most up-sampling operators, DySample mainly adopts the point resampling method instead of the traditional kernel-based method, which achieves efficient and flexible up-sampling operation by generating up-sampling locations, using a fixed bilinear kernel, reducing computational complexity and resource consumption, and relying on a smaller number of feature map channels.

Figure 6a shows the DySample sampling-based dynamic up sampling schematic. The input feature (X) is processed by the sample point generator, which is responsible for creating a set of sample points with two channels to form the sample data set (S).

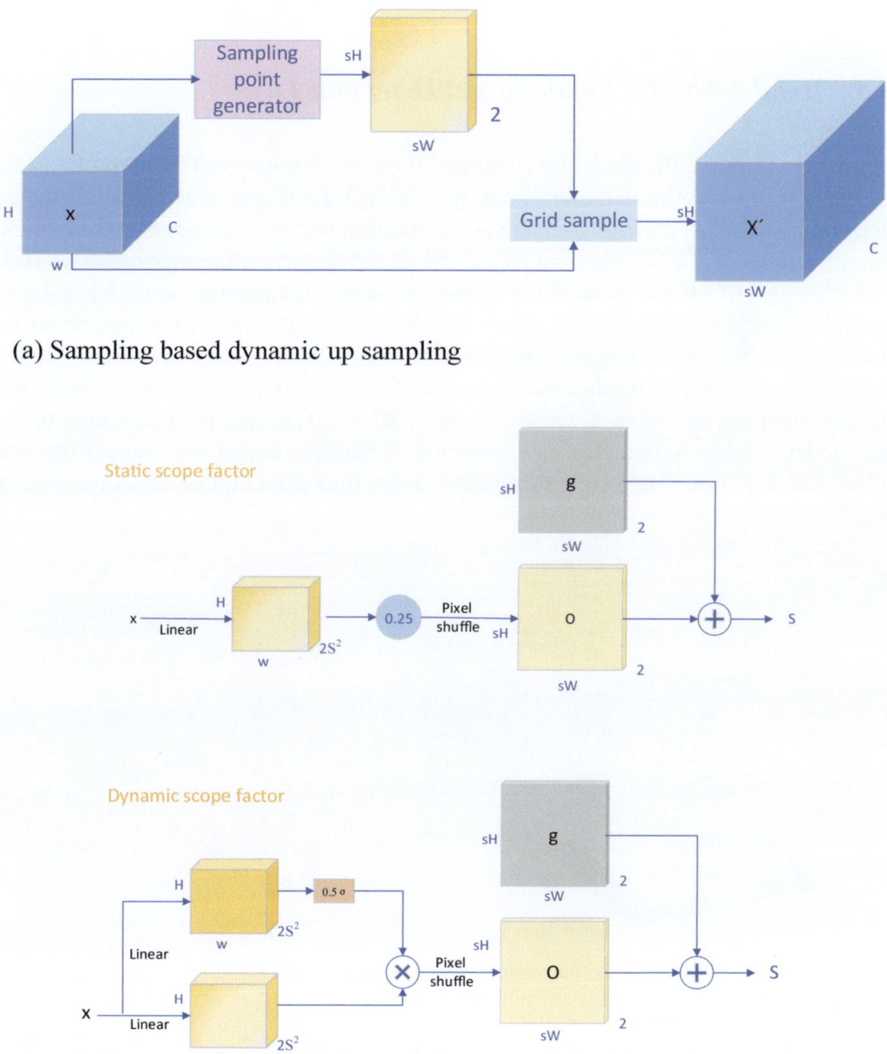


(a) Sampling based dynamic up sampling

(b) Sampling point generator in Dysample

**Fig. 6** Structural diagram of DySample

Then, the data set S and the input data X are resampled using the grid sample function (grid sampling method) to obtain the up-sampling feature (X′). Figure 6b shows the working principle of the sampling point generator based on two different range factors. Both range factors are combined with the range factors and then generated by pixel shuffling technology to generate offset (O), and then added with the original grid position (g) to obtain the sample set (S). In the dynamic range factor σ is a Sigmoid function, which is used to generate the dynamic range factor.

## 3.3 Head Adding a Small Target Detection Layer

In order to improve the detection accuracy of the model, inspired by literature [23], we have improved the detection head part of YOLOv8 and added a small target detection head P2. This improvement is the addition of detection heads in the feature pyramid network (FPN) structure, especially on the output feature map of 160 × 160. The core idea of FPN is to combine high-level semantic information with low-level spatial information through top-down feature fusion, thereby achieving effective detection of multi-scale targets. Such a design is particularly suitable for small target recognition because it retains more detailed features. At the same time, there are defects such as off-centered, which retains a 20 × 20 detection head, which helps the model to adapt to complex environments of multiple scales and reduces the risk of missed detection. Therefore, this paper adopts four detection heads. As shown in Fig. 7.

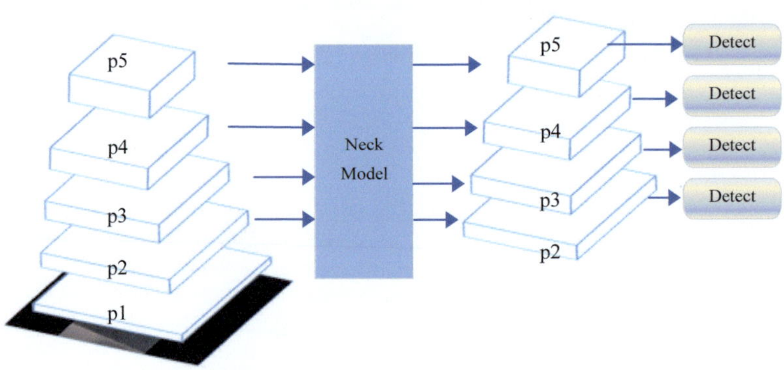


**Fig. 7** Schematic diagram of the four-detector head model

### 3.4 Unsupervised Learning Model

Although supervised learning methods have achieved remarkable results in quick plug connector defect detection, they rely on large amounts of labeling data, while defect samples are sparse and labeling costs are high in actual production. Unsupervised learning, as a method of learning features from data without labeling data, provides new ideas for solving this problem. This section will introduce several commonly used unsupervised models.

Autoencoder (AE) [24] compresses the input data into a low-dimensional representation through the encoder, and then reconstructs the original data through the decoder. Its goal is to minimize reconstruction errors and thus learn effective feature representations of data. The autoencoder performs excellently in the abnormality detection task and is able to mark samples that deviate from the normal distribution as abnormal by learning the characteristic distribution of normal samples.

Generative Adversarial Networks (GAN) [25] consists of a generator and a discriminator.

The generator tries to generate samples similar to the real data, while the discriminator tries to distinguish between real samples and generated samples. Through adversarial training, GANs are able to generate high-quality samples that can be used for data augmentation or anomaly detection.

Contrastive Learning [26] learns the feature representation of the data by maximizing similarity between similar samples and differences between dissimilar samples. Its core idea is to generate positive sample pairs through data augmentation and optimize the model through comparison loss. Comparative learning performs well in unsupervised feature extraction, especially for small sample scenarios.

Clustering algorithms (such as K-Means [27]) discover the intrinsic structure of the data by dividing the data into several clusters. In defect detection, clustering algorithms can be used to group unlabeled data to discover new defect categories.

Variational Autoencoder (VAE) [28] generates new samples and detects exceptions by learning the probability distribution of data.

Compared with autoencoders, VAE can generate more diverse samples, suitable for scenarios with complex data distribution. In the above method, the AE can effectively learn the feature distribution of normal samples by reconstructing error detection, but it is poorly adaptable to multi-scale defects and is more sensitive to complex background interference. GANs require a large amount of data to generate high-quality pictures. Comparative learning uses maximizing the similarity of similar samples to significantly improve the model sensitivity to small-target defects, but it has strong dependence on data enhancement strategies and high computing costs. Clustering algorithms (such as K-Means) discover potential defect categories by grouping unlabeled data, which is simple and efficient, but sensitive to the initial clustering center and difficult to deal with defects of varying sizes. VAE generates diverse samples by learning the data probability distribution, which is suitable for complex data distribution, but its reconstruction accuracy is low, which may affect the detection effect of small target defects. Comprehensive consideration, combined

**Table 2** Experimental environment configuration for deep learning

| Part | Configuration |
|---|---|
| Operating system | Ubuntu 20.04 |
| Compilers | Python 3.9.18 |
| Deep learning frameworks | Pytorch 1.13.1 |
| CPU | Intel(R) Xeon(R) Gold 6149 CPU @ 3.10 GHz |
| GPU | NVIDIA TITAN Xp × 2 |
| CUDA | CUDA11.6 |

**Table 3** Hyperparameter configuration for deep learning

| Parameter | Meaning | Value |
|---|---|---|
| Img-size | The size of input image | 640 × 640 |
| Batch size | Number of images processed at a time | 8 |
| Epochs | Number of trains | 400 |
| lro | Initial learning rates | 0.01 |
| lrf | Learning rate attenuation factor | 0.01 |
| Weight decay | Parameters for each decay of weights | 0.0005 |
| Momentum | Parameters for delayed gradient descent | 0.937 |

with AE and comparative learning mechanisms, an unsupervised model that takes into account both accuracy and efficiency.

# 4 Experiment Detail

## *4.1 Experimental Environment and Parameter Configuration*

The experimental environment configuration of the QCF-YOLO model is shown in Table 2 and the hyperparameter configuration is shown in Table 3.

## *4.2 Experiment Platform*

The experimental platform for the QCF testing is constructed, as shown in Fig. 8. In order to ensure the image quality, this paper adopts the ring light source, and the brightness of the light source is controlled by the light source controller, so as to capture the image with high quality. The QCF is placed on the fixture for fixing, and the data is collected by the camera, transferred to the control board for processing, and

**Fig. 8** Experimental platform for testing

then the predicted results are shown on the display. Due to the fixed-point inspection, manual operation is still required to replace parts.

# 5 Analysis of Datasets

## *5.1 Ablation Experiment*

In order to verify the effectiveness of each module in the QCF-YOLO network for the detection of surface defects of the quick plug connector, this paper improves based on the YOLOv8 model and performs ablation experiments on each module, using P, R, AP, mAP50, Params, FLOPs and FPS as Evaluation indicators. The experimental results are shown in Table 4.

Solution B uses C3GNet network. Although the accuracy of cracking defects has decreased by 2%, Params and FLOPs have decreased by 20.1 and 16.2%, respectively, significantly improving efficiency. By reducing the number of channels, Params and

**Table 4** Ablation experiments

| Number | Method | mAP50/% | Params/106 | FLOPs/G | FPS/s |
|---|---|---|---|---|---|
| A | YOLOv8 | 89.9 | 2.791 | 7.4 | 82.2 |
| B | A + C3GNet | 89.8 | 2.230 | 6.2 | 89.7 |
| C | A + S_FPN | 89.7 | 1.893 | 5.2 | 95.3 |
| D | A + DySample | 92.8 | 3.019 | 8.1 | 78 |
| E | A + p2 | 97.4 | 2.804 | 11.9 | 77.3 |
| F | B + S_FPN | 86.9 | 1.184 | 3.3 | 90.3 |
| G | F + DySample | 91.1 | 1.903 | 5.2 | 86.8 |
| H | G + p2 | 95.5 | 1.187 | 6.1 | 69.8 |

FLOPs were reduced by 32.2 and 29.7%, respectively, and FPS was increased to 95.3 frames, focusing on real-time. Solution D uses the DySample up sampling operator, and mAP50 is increased by 2.9%, but Params and FLOPs are increased by 8.1 and 9.5% respectively, and FPS is reduced to 78 frames, which is suitable for scenarios with high accuracy requirements. Solution E: The mAP50 was increased by 7.5% by adding P2 detection heads, but the FLOPs increased by 60.8% and the FPS dropped to 77.3 frames, mainly improving the detection accuracy. Solution F combines C3GNet and S_FPN, although mAP50 has decreased by 2.9%, Params and FLOPs have decreased by 46.9 and 46.7%, respectively, improving computing efficiency. Solution G added DySample based on F, and mAP50 increased by 4.2%, but Params and FLOPs increased by 60 and 57% respectively. Solution H adds P2 detection head on G, mAP50 increases by 4.4%, Params decreases by 57.5% compared with A, and FLOPs increases by 17.6%, which is suitable for high-precision needs.

## 5.2 *Fusion Experiment of DySample*

Since there are three up sampling operations in this article, in order to find the best embedding location for DySample, this article conducts a fusion experiment of DySample. In order to control variables and ensure the reliability of the experiment, the fusion experiment was conducted under experimental conditions where the backbone network is C3Gnet, the neck network is S_FPN, and the head network is 4 detection heads. The results of the fusion experiment are shown in Table 5 respectively represent the three up sampling positions, and "√" means that the up-sampling operation is used here using the DySample up sampling operator.

It can be seen from Table 5 that compared with the C3GNet_P2 model, DySample up sample operator is embedded in three different locations, and there are 7 embedding schemes in total. Different solutions are used to detect models differently. Among them, the mAP50% of solution 8 has the highest, while Map50% increased by 1.8% compared with the C3GNet_P2 model; the mAP50% of solution 2 is the

**Table 5** Fusion experiment of DySample

| Number | C3Gnet_p2 | 1 | 2 | 3 | P/% | R/% | mAP50/% |
|---|---|---|---|---|---|---|---|
| 1 | √ | | | | 92.6 | 87.2 | 93.7 |
| 2 | √ | √ | | | 91.8 | 85.2 | 93 |
| 3 | √ | | √ | | 94.2 | 88.2 | 95 |
| 4 | √ | | | √ | 94.2 | 88.8 | 95.1 |
| 5 | √ | √ | √ | | 94.1 | 87.9 | 94.5 |
| 6 | √ | √ | | √ | 92.7 | 87.6 | 94.1 |
| 7 | √ | | √ | √ | 93.8 | 86.1 | 93.8 |
| 8 | √ | √ | √ | √ | 94.6 | 89.1 | 95.5 |

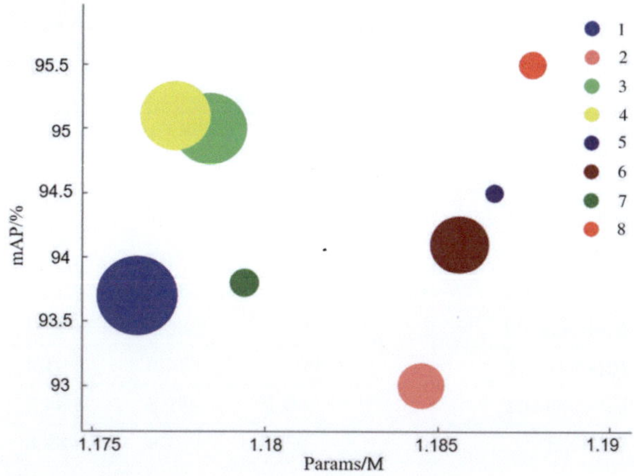


**Fig. 9** Comparison of model performance

lowest-precision solution among all solutions, with Map50% being 93%. Figure 9 shows a performance comparison diagram of all solutions. The horizontal axis represents the size of the model parameter quantity, the vertical axis represents the mAP50 value, and the size of the point represents the model processing speed. The larger the area, the faster the processing speed. It can be seen from Fig. 9 that the mAP 50% value of solution 8 has the highest value, but its parameters are slightly larger and the processing speed is slightly slower. However, taking into account comprehensively, it is believed that Scheme 8 is the best effect, that is, the DySample up sampling operator is embedded in all three positions.

## *5.3 Contrast Experiment*

Firstly, in order to verify the validity of the backbone network (C3GNet) of the QCF-YOLO model, different backbone networks are selected on the homemade dataset and compared with the method of this paper. The experimental results are shown in Table 6.

When the Backbone network is Efficientvit2, the mAP50 value is the highest, reaching 94.7%, but its parameter quantity and floating-point number operation (GFLOPs) are also the largest. When Backbone is C3GNet, the parameter quantity is the smallest, only 2.299 M, which is equivalent to 10% of the maximum parameter quantity. Although C3GNet mAP50 is 89.8%, a decrease of 2.1% compared to the original network, the number of parameters is reduced by 23.5%. When the Backbone network is Efficientvit1, the model GFLOPs value is the smallest, that is, the inference speed is the fastest, and the mAP50 value is high, reaching 91.4%. However, compared with C3GNet, the parameter volume of Efficientvit1 is larger, an increase of 214%, a decrease of 0.5% compared with the original network mAP50 value, but the parameter volume is increased by 141%. Taking into account various indicators

**Table 6** Comparative experiments on lightweight backbone networks

| Backbone | Neck | P (%) | R (%) | Map50 (%) | GFLOPs (G) | Params (106) |
|---|---|---|---|---|---|---|
| YOLOv8 | FPN | 94.3 | 84.1 | 91.9 | 8.1 | 3.007 |
| ShuffleNet V1 | | 90.7 | 82.2 | 89.4 | 6.5 | 3.477 |
| ShuffleNet V2 | | 90.9 | 83.2 | 89.9 | 7.4 | 2.791 |
| Mobilenext1 | | 93.9 | 85.4 | 92 | 15.7 | 6.065 |
| Mobilenext2 | | 94 | 85.5 | 92.1 | 10.1 | 3.758 |
| Mobilenext3 | | 93.3 | 86.9 | 92.3 | 9.5 | 5.660 |
| Efficientvit1 | | 94 | 84.9 | 91.4 | 5.6 | 7.237 |
| Efficientvit2 | | 96.1 | 89.3 | 94.7 | 52.1 | 21.752 |
| C3GNet | | 93.7 | 81.9 | 89.8 | 6.2 | 2.299 |

comprehensively, although the mAP50 value of C3GNet is relatively low, its parameters are small, bringing significant computing and storage advantages. This allows C3GNet to maintain high accuracy while having high fault tolerance and efficiency, confirming the effectiveness of C3Gnet.

In order to improve the resolution and quality of the feature map and retain more edge feature information, this paper combines different up sampling operators with the QCF-YOLO model and embeds them into three positions of the model, and the three positions are embedded in the same position up sampling operator. In the selection of up sampling operators, the original up sampling operator of YOLOv8 (i.e. nearest neighbor interpolation), the widely used CARAFE, and the DySample up sampling operator first proposed in YOLOv9 are mainly considered. The specific experimental results are shown in Table 7.

As can be seen from Table 7, when using CARAFE, GFLOPs rose by 16.4%, Params rose by 16.7%, but mAP50 rose by 1.6%. Accuracy P and recall R rose by 3.2 and 1.6% respectively; when using DySample, GFLOPs remained basically the same, Params rose by 0.9%, but mAP50 rose by 1.8%. Accuracy rate P and recall rate R increased by 2 and 1.9% respectively. Compared with CARAFE, using DySample can only increase slightly while ensuring the accuracy improvement, the number of parameters increases slightly, the floating-point number operation GFLOPs is relatively reduced, and the model complexity is reduced. Therefore, this article uses DySample as the upsampling operator, which can effectively improve the model accuracy and only increase the number of parameters.

**Table 7** Comparison of results of different dynamic up samplers

| Model | Module | P/% | R/% | mAP50/% | GFLOPs/G | Params/106 |
|---|---|---|---|---|---|---|
| QCF-YOLO | Up sample | 92.6 | 87.2 | 93.7 | 6.1 | 1.176 |
| | CARAFE | 95.8 | 88.8 | 95.3 | 7.1 | 1.373 |
| | DySample | 94.6 | 89.1 | 95.5 | 6.1 | 1.187 |

In order to evaluate the performance of the QCF-YOLO network proposed in this paper on the QCF detection, the method QCF-YOLO in this paper is compared with SSD, Faster R-CNN, Yolov3, Yolov5s, Yolov6, Yolov7, Yolov8, Yolov8s, and Yolov9, and the experimental conditions are consistent. The specific experimental results are shown in Table 8, and Fig. 10 shows the comparison of experimental results of mainstream models.

Through the analysis of Table 8, it can be seen that compared with SSD and Faster R-CNN, although SSD performs well in detection accuracy, the recall R and mAP50 become lower, the model parameters are large, and the model complexity is higher. FasterR-CNN performs well in recall R, but the Map50 value is low, and the complexity of GFLOPs and model parameter volume Params are higher than

**Table 8** Results of the mainstream model test comparison table

| Model | P/% | R/% | mAP50/% | GFLOPs/G | Params/106 | FPS /s |
|---|---|---|---|---|---|---|
| SSD | 98.5 | 28.2 | 57.92 | 64.75 | 28.26 | 56.9 |
| Faster R-CNN | 50.8 | 85.18 | 75.9 | 401.78 | 136.77 | 22 |
| YOLOv3 | 67.5 | 40.6 | 49.6 | 155.33 | 61.545 | 24.7 |
| YOLOv5s | 92.23 | 85.35 | 91.46 | 15.80 | 7.024 | 15.8 |
| YOLOv6 | 95.32 | 85.75 | 91.11 | 156.01 | 54.209 | 18.2 |
| YOLOv7 | 89.2 | 83 | 88.77 | 104.70 | 36.9 | 20.2 |
| YOLOv8 | 94.4 | 84.2 | 91.9 | 8.10 | 3.007 | 82.2 |
| YOLOv8s | 97.17 | 90.64 | 96.05 | 28.80 | 11.167 | 97.1 |
| YOLOv9 | 95.8 | 90.3 | 95.1 | 263.90 | 6.051 | 75 |
| YOLOv10n | 95.4 | 92 | 96.3 | 6.7 | 2.696 | 97.9 |
| RT-DETR | 86.2 | 86.9 | 92.6 | 108.3 | 32.816 | 14.5 |
| QCF-YOLO | 94.6 | 89.1 | 95.5 | 6.10 | 1.187 | 69.8 |

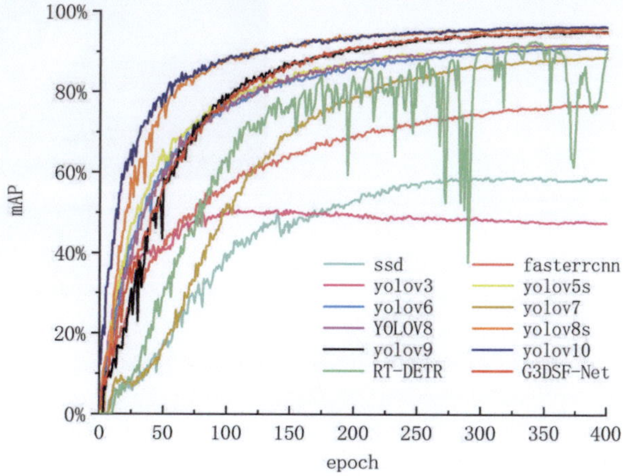


**Fig. 10** Comparison of detection results of different model

other models. RT-DETR, the method in this paper performs better than SSD, Faster R-CNN and RT-DETR models in terms of detection accuracy P. Compared with the mainstream YOLO series models, the mAP50 value of Yolov10 is the highest, but the Params and GFLOPs of the model are 9.8% and 1.27 times higher than QCF-YOLO, respectively, and the mAP50 value is only 0.8%. From Fig. 10, it can be seen that during the entire training process, the mAP50 value of QCF-YOLO is smaller than that of YOLOv9. The mAP50 values of YOLOv8s and YOLOv10 are slightly higher than those of QCF-YOLO, but the model parameters of YOLOv8s are large. More complex and requires more resources. Compared with QCF-YOLO, the YOLOv10n model has a large number of parameters.

To sum up, QCF-YOLO performs well in model detection accuracy, model complexity and model parameters, and has the best overall performance. The effect of QCF-YOLO model detection is shown in Fig. 11.

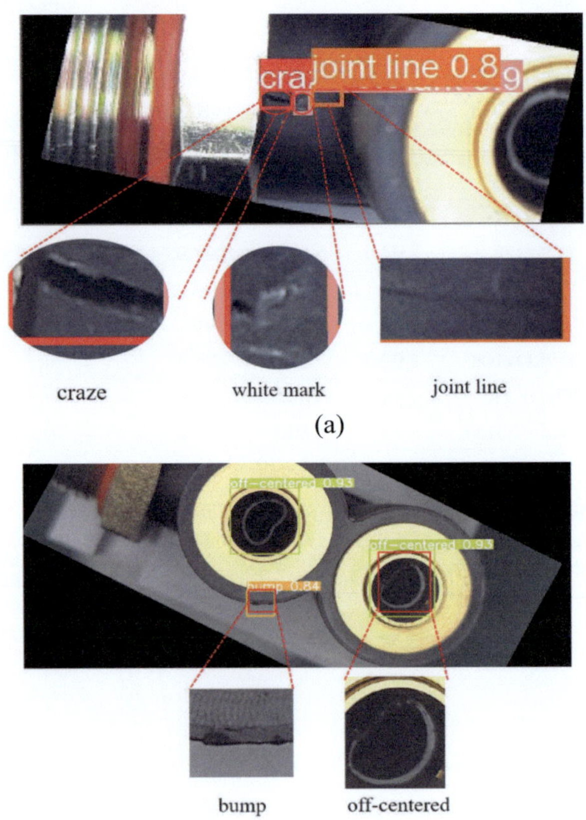


**Fig. 11** QCF-YOLO model defect detection results

## 6 Conclusion and Discussion

In response to the edge deployment and defect detection requirements of fast plug connector models, this paper proposes a lightweight QCF-YOLO model. The model mAP50 reaches 95.5%, the frame rate is 69.8FPS, the parameter quantity is 1.187 M, and the GFLOPs is 6.1. The specific conclusions are as follows:

The lightweight C3GNet backbone network and S-FPN feature fusion network are adopted, and the model parameters are reduced by 57.6% compared with YOLOv8, and the accuracy loss is only 3%. The DySample up sampling operator was introduced to further reduce the parameter volume by 31.8%, while improving the accuracy by 1.2%. Adding p2 small target detection head significantly improves the detection capability of micro defects and improves accuracy by 3.6%, meeting real-time detection needs.

Through analysis, we will design an unsupervised model combining autoencoder and contrast learning mechanism. The next work mainly studies the feasibility of an unsupervised model solution combining autoencoder and contrast learning mechanism, and realizes the coordinated cooperation of; cloud-edge-end collaborative architecture; to improve the degree of detection automation.

The current study has revealed certain limitations in addressing the small-sample problem. While the primary strategy employed involved expanding the dataset through data augmentation techniques, this approach may exhibit constrained effectiveness when confronted with more complex detection environments. Future research will prioritize the enhancement of model generalization capabilities to ensure robust adaptation across diverse and challenging detection scenarios.

## References

1. Lee, J., Hong, I., Kim, S.J.K.: Experimental study of brake pipe oil leak by sealing performance improvement. Trans. Korean Soc. Automot. Eng. **21**, 819–824 (2013)
2. Lu, Y., Zhu, Q., Guo, B.: The study of automobile brake chamber performance test system. Appl. Mech. Mater. **543–547**, 1286–1290 (2014)
3. Wang, Y., Yao, Q., Kwok, J.T., Ni, L.M.: Generalizing from a few examples: a survey on few-shot learning. ACM Comput. Surv. **53**(3), 1–34 (2020)
4. Qiu, Y., Lu, J.: A visualization algorithm for medical big data based on deep learning. Measurement **183**, 109808 (2021)
5. Zhang, Z., Zheng, Z., Bai, Y.: A new method for detecting defect of annular joint of automotive brake fluid pipe. Machinery **61**(10), 63–69 (2023)
6. Zhu, S., Xia, X., Zhang, Q., Belloulata, K.: An image segmentation algorithm in image processing based on threshold segmentation. City (2007)
7. Zeng, N., Wu, P., Wang, Z., Li, H., Liu, W., Liu, X.: A small-sized object detection oriented multi-scale feature fusion approach with application to defect detection. IEEE Trans. Instrum. Meas. **71**, 1–14 (2022)
8. Yu, T., Chen, W., Junfeng, G., Poxi, H.: Intelligent detection method of forgings defects detection based on improved EfficientNet and memetic algorithm. IEEE Access **10**, 79553–79563 (2022)

9. Peng, Y., Wang, Z., Zhang, Q.: Application of improved Faster-RCNN algorithm in internal defect detection of polyethylene pipe joint. J. Appl. Acoust. **42**(5), 984–992 (2023)
10. Ren, S., He, K., Girshick, R., Sun, J.: Faster R-CNN: towards real-time object detection with region proposal networks. IEEE Trans. Pattern Anal. Mach. Intell. **39**(6), 1137–1149 (2017)
11. Ge, D., Li, H., Zhang, L., Liu, R., Shen, P.: Survey of lightweight neural network. J. Softw. **31**(9), 2627–2653 (2020)
12. Dahai, L., Zhihui, C., Xianqi, L., Qi, Z., Nanxing, W.A: Lightweight convolutional neural network for recognition and classification for $Si_3N_4$ chip substrate surface defects. Ceram. Int. **49**(22), 35608–35616 (2023)
13. Luo, Y., Ci, Y., Jiang, S., Wei, X.: A novel lightweight real-time traffic sign detection method based on an embedded device and YOLOv8. J. Real-Time Image Process. **21**(2), 24 (2024)
14. Zhang, J., Jing, J., Lu, P., Song, S.: Improved MobileNetV2-SSDLite for automatic fabric defect detection system based on cloud-edge computing. Measurement **201**, 111665 (2022)
15. Shi, W., Cao, J., Zhang, Q., Li, Y., Xu, L.: Edge computing: vision and challenges. IEEE Internet Things J. **3**(5), 637–646 (2016)
16. Shuai, Y., Lin, Z., Chen, W., Shenghuai, W., Yu, T.: SF-YOLO: an evolutionary deep neural network for gear end surface defect detection. IEEE Sens. J. **24**(13), 21762–21775 (2024)
17. Xi, L., He, D., Liu, H.: PRAAD: pseudo representation adversarial learning for unsupervised anomaly detection. J. Inform. Secur. Appl. **89**, 103968 (2025)
18. Kozamernik, N., Bračun, D.: A novel FuseDecode Autoencoder for industrial visual inspection: incremental anomaly detection improvement with gradual transition from unsupervised to mixed-supervision learning with reduced human effort. Comput. Ind. **164**, 104198 (2025)
19. Tao, X., Cao, Y., Jiang, Y., Wu, X., Yan, D., Xue, W., Zhuang, S., Yang, X., Huang, R., Zhang, J., Ni, D.: Enhancing lesion detection in automated breast ultrasound using unsupervised multi-view contrastive learning with 3D DETR. Med. Image Anal. **101**, 103466 (2025)
20. Song, G., Hwang, J., Kim, J., Kim, H., Lee, S., Park, J., Cho, G.: Automatic classification of multi-channel PSD results combining unsupervised and supervised learning. Nucl. Eng. Technol. **103331** (2024)
21. Han, K., Wang, Y., Tian, Q., Guo, J., Xu, C., Xu, C.: GhostNet: more features from cheap operations. City (2020)
22. Liu, W., Lu, H., Fu, H., Cao, Z.: Learning to upsample by learning to sample. City (2023)
23. Chen, X., Jiang, Z., Piao, Y., Yang, J., Zheng, H., Yang, H., Chen, K.: SF-Yolov8n: a novel ultralightweight and high-precision model for detecting surface defects of dental nails. IEEE Sens. J. **24**(12), 20103–20113 (2024)
24. Li, P., Pei, Y., Li, J.: A comprehensive survey on design and application of autoencoder in deep learning. Appl. Soft Comput. **138**, 110176 (2023)
25. Krichen, M.: Generative adversarial networks. City (2023)
26. Lei, Y., Jing, R., Li, F., Gao, Q., Deng, C.: A transformer-based dual contrastive learning approach for zero-shot learning. Neurocomputing **626**, 129530 (2025)
27. Win, N.S.S., Li, G., Lin, L.: Revolutionizing early breast cancer screening: advanced multi-spectral transmission imaging classification with improved Otsu's method and K-means clustering. Comput. Biol. Med. **184**, 109373 (2025)
28. Sado, I.T., Fitime, L.F., Pelap, G.F., Tinku, C., Meudje, G.M., Bouetou, T.B.: Early multi-cancer detection through deep learning: an anomaly detection approach using Variational Autoencoder. J. Biomed. Inform. **160**, 104751 (2024)

# Designing an Origami-Inspired Soft-Actuated Smart System for Portable Pelvic Rehabilitation

**Alessia Buffagni, Elisa Venturini, and Giovanni Borga**

**Abstract** This paper presents the study of an origami-inspired soft-actuation smart system designed for pelvic floor rehabilitation. Initially conceived for women who have undergone female genital mutilation (FGM) and deinfibulation, the system extends its applicability to a broader user base, addressing diverse pelvic floor dysfunctions. Starting from an in-depth socio-medical analysis of FGM and its long-term consequences, this research adopts an integrated and multidisciplinary approach that combines smart healthcare solutions with soft robotic systems. The study explores the potential of origami-inspired soft actuation to enhance rehabilitation effectiveness by ensuring adaptability, personalized treatment, and comfort. The proposed system incorporates an inflatable actuator capable of dynamic and controlled interaction with the pelvic floor muscles, facilitating both active and passive training. By leveraging sensor integration and data monitoring, the device enables precise adjustments tailored to individual patient needs, supporting remote therapy management and telemedicine applications. This work contributes to the development of innovative rehabilitation solutions by bridging medical, technological, and cultural considerations, ultimately promoting accessible and patient-centered approaches in pelvic health.

**Keywords** Soft-actuated system · Smart healthcare · Pelvic rehabilitation

A. Buffagni (✉) · E. Venturini · G. Borga
Università Iuav Di Venezia, Venezia, Italy
e-mail: abuffagni@iuav.it

E. Venturini
e-mail: e.venturini@stud.iuav.it

G. Borga
e-mail: borga@iuav.it

 
Y. Ma (ed.), *Proceedings of 4th International Conference on Intelligent Systems Design and Engineering Applications*, Lecture Notes in Electrical Engineering 1478,
https://doi.org/10.1007/978-981-95-1802-9_11

## 1 Introduction

According to a joint definition by World Health Organisation (WHO), United Nations International Children's Emergency Fund (UNICEF) and United Nations Fund for Population Activities (UNFPA) in April 1997, Female Genital Mutilation (FGM) includes all procedures involving the partial or total excision of the external female genitalia or the infliction of other injuries or disfigurement for cultural or non-medical reasons [1].

The motivations behind FGM are complex and range from cultural to sociological factors such as beliefs about purity, social integration, socio-economic pressures related to marriage and women's status, control of female sexuality and aesthetic ideals. Despite being classified as a human rights violation by both the United Nations (UN) [1] and WHO, FGM continues to be practiced in many countries, with profound effects on the physical, sexual and mental health of the women affected, as well as significant social repercussions on the from birth to adulthood.

The age at which girls and young women are subjected to these procedures varies across cultural groups, often occurring during the pre-pubertal years [2]. The practice of FGM can lead to serious long-term complications, putting the life of the victim at risk. Immediate health consequences can include excessive bleeding, shock, infection, HIV transmission, urinary retention and extreme pain [3]. Survivors often require urgent medical intervention, including surgery to stop bleeding, antibiotics for infections, and emergency obstetric care for complicated deliveries [2, 4].

FGM is typically conducted in unsafe domestic environments, using rudimentary and brutal tools. Over time, the psychological trauma can mutate into anxiety and depression, with consequences for daily and social life. In the psychosexual sphere, the consequences can manifest themselves in involuntary contractions of the vulvo-perineal muscles, which can hinder or prevent penetration, even under medical supervision [3].

A careful medical analysis of patients experiencing muscular changes due to inappropriate use of the pelvic and urethral muscles has shown that restoring the dilation and contraction functions of the vaginal canal and rehabilitating the pelvic musculature is a fundamental step in the management of postoperative treatment after reopening the vaginal canal (deinfibulation).

Starting from an in-depth socio-medical analysis of FGM, this article presents a wide-ranging project that supports an integrated and multidisciplinary approach combining smart healthcare and soft systems. The project focuses on the development of a portable device for physiotherapeutic rehabilitation of the pelvic floor, combined with an integrated information system that supports the user throughout the physical and psychological therapeutic process.

With the help of telemedicine, remote therapy outcomes provide feedback to the referring doctor, enabling them to adapt and refine the type of treatment according to the needs and the problem to be treated, whatever the primary cause.

The use of soft actuators also makes it possible to act on the treated area in an adaptable and responsive manner, ensuring accurate feedback and improving the effectiveness of the therapy, while maintaining complete safety. The project's versatility demonstrates its applicability not only to the issue analyzed in the article but also to the treatment of various pathologies affecting normal pelvic floor functions, making it suitable and accessible to a broader range of patients.

# 2 Method

The research unfolds through several phases. Initially, a comprehensive analysis of the state of the art in pelvic floor rehabilitation is conducted, mapping existing tools and technologies, classifying devices (inflatable devices, vaginal dilators, sex toys), and identifying key functional and technological gaps.

Subsequently, the study focuses on inflatable devices, exploring their advancements in medical applications. In parallel, user needs are identified through interviews with patients and healthcare professionals (not discussed in this article), along with the technical requirements for integrating technologies such as pressure sensors and inflatable actuators to ensure comfort and safety. This data guides the device's design process: the development of an origami-inspired inflatable actuator, the selection of electromechanical components and biocompatible materials, and the design of an anatomical shell.

## *2.1 Current Tools for Pelvic Rehabilitation*

To date, no devices are specifically designed for pelvic floor rehabilitation to address conditions resulting from infibulation. However, the complications caused by infibulation often share similarities with other pelvic floor pathologies, making it possible to adapt existing devices for therapeutic purposes. These devices can generally be classified into three categories: inflatable devices, vaginal dilators, and sex toys.

**Inflatable Devices**. Inflatable devices encompass all instruments designed to increase their initial volume through the injection of air or fluids. In gynecology, these devices are often employed to support the re-education of the pelvic muscular system by creating resistance for patients to work against. Among the most frequently used tools in this category are pelvic floor rehabilitation probes. These devices utilize contraction training techniques to strengthen the pelvic floor and are effective in addressing disorders affecting the genito-urinary and colorectal systems [5].

**Vaginal Dilators**. Vaginal dilators are medical instruments specifically designed for gradual vaginal dilation. Most are constructed from medical-grade silicone and feature a concave opening that facilitates easier insertion without requiring an applicator for leverage. These dilators are typically available in sets with progressively increasing diameters, enabling incremental dilation and adaptation over time. This functionality supports therapeutic applications such as the treatment of vaginal stenosis and other pelvic floor dysfunctions [6]. For instance, a recent study proposed inflatable vaginal dilators for treating radiotherapy-induced vaginal stenosis [7].

**Sex Toys**. Sex toys are frequently incorporated into physiotherapy protocols due to their anatomical design and supplementary functions, which provide varied stimuli for the re-education of the pelvic floor muscles. These devices contribute to restoring proper urinary and sexual function in women. However, the cultural perception of sex toys presents a significant barrier to their use. Many women associate these devices exclusively with the sexual domain, often considering them culturally inappropriate or morally impure, which limits their acceptability as medical tools [8].

## 2.2 *Limitations of the Current Tools and Need for Targeted Solutions*

While these existing devices offer some therapeutic benefits, none are specifically designed to address the unique challenges associated with infibulation-related pelvic floor dysfunction. This highlights the urgent need for innovative, culturally sensitive, and purpose-built solutions to effectively rehabilitate pelvic floor muscles while addressing the physical, psychological, and cultural complexities involved.

In this context, soft actuation emerges as an ideal solution for pelvic floor rehabilitation, thanks to its adaptability, precision, and capacity to address cultural sensitivities. Soft actuators are flexible, allowing them to conform to the complex anatomy of individuals affected by infibulation-related dysfunction. Unlike rigid devices, they provide personalized support while minimizing discomfort and reducing the risk of injury, ensuring a patient-specific and effective therapeutic intervention.

Additionally, soft actuation allows for precise control of applied forces, enabling the device to dynamically respond to the patient's movements or contractions. This controlled interaction enhances rehabilitation efficacy by providing the appropriate resistance or assistance during therapy [9], making it particularly beneficial for re-educating weakened or damaged pelvic floor muscles.

Given the physical and psychological trauma associated with infibulation, soft actuators offer a gentle, non-invasive approach that reduces discomfort and addresses the needs of patients with heightened sensitivity or scar tissue. They can also incorporate sensors to provide real-time feedback on muscle activity and therapy progress, enabling remote monitoring and therapy adjustments via telemedicine. The discreet design and neutral appearance of soft-actuated devices address cultural sensitivities and improve acceptance among patients and their communities. By operating within safe limits, these devices reduce the risk of overexertion or unintended injury, which is particularly important for individuals recovering from complications related to infibulation. Moreover, the versatility of soft actuation technology extends its applicability beyond infibulation-related conditions, making it suitable for a wide range of pelvic floor dysfunctions. This adaptability increases accessibility and enhances therapeutic outcomes for diverse patient populations. Soft actuation also offers a transformative approach to pelvic floor rehabilitation, combining flexibility, precision, safety, and cultural sensitivity. These qualities position it as a powerful solution for addressing the complex physical, psychological, and cultural challenges associated with pelvic health issues, while also enabling broader applications in clinical practice.

## *2.3 Advancements in Soft Actuators for Biomedical Applications*

Soft-actuation enables safer human-technology interaction making it suitable for biomedical rehabilitation and assistance [10, 11]. As identified by Pan et al. [10] various types of soft actuators have been developed to address specific needs in biomedical rehabilitation and wearable robotics. The scientific literature showcases a wide range of applications, including solutions tailored for spine [12], knee [13] hand [14], neurological recovery and post-stroke motor function enhancement [15], gait training [16, 17], and even in post-fall postural re-education in elderly women [18]. These examples highlight the versatility of soft actuation in addressing diverse musculoskeletal and neuromuscular conditions.

**Geometrically Responsive Actuators that Contract and Expand (GRACE).** Biomimetic machines, capable of integrating into natural and social environments, have a wide range of applications, from the conservation of biodiversity to the daily care of the elderly. Although artificial actuators have achieved performance similar to that of human muscles in terms of contraction, their versatility and fluidity of movement, typical of complex muscle arrangements, remain largely unmatched. The scientific literature introduces a new class of pneumatic actuators called GRACE (Geometrically Responsive Actuators that Contract and Expand) [19], which are based on geometric principles for contraction and extension. GRACEs consist of a single folded diaphragm and do not require any deformation-limiting elements. They can contract and extend according to a mathematical model and can be manufactured in different sizes and materials, offering variable mechanical performance and enabling a wide range of realistic movements. GRACEs can be produced at low cost and even integrated directly into functional devices. This simplifies and accelerates the prototyping and production of devices based on artificial muscles [20]. They can be designed to maximise contraction, extension or antagonism, and can be manufactured with different materials and low-cost 3D printers in various sizes.

Finally, they can be easily integrated into complex devices that require sophisticated and diverse movements. All of these features are possible thanks to the special structure of GRACEs: a low-stress geometric transformation of the actuators, in which the folds are designed to allow contraction and/or extension without the need for constraining components such as traction-resistant fibres or rigid end pieces (see Fig. 1).

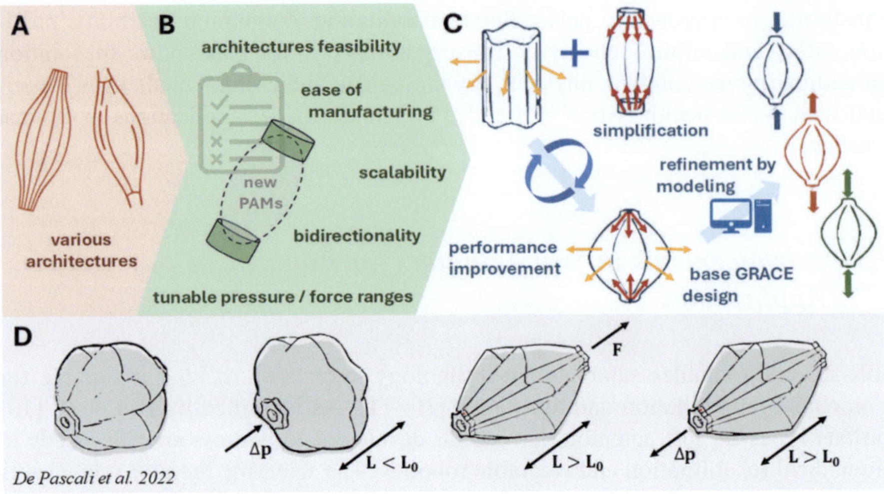


**Fig. 1** Concept of GRACE artificial musculature developed by De Pascali et al. [19]

**Origami-inspired inflatable actuators**. In recent years, origami-based inflatable actuators have gained considerable attention due to their advantages, such as strength, efficiency, and adaptability. Innovations in origami-inspired design have opened new possibilities for soft robotic systems, particularly in medical applications [21, 22]. Origami-inspired soft actuators for rehabilitation represent a valuable and rapidly evolving area of research at the intersection of design, materials, robotics, and healthcare. These actuators leverage the principles of origami to create structures and mechanisms that are lightweight, flexible, adaptable, and capable of complex movements—examples include origami actuators for wrist brace [22], origami-inspired soft actuated assistive exoskeleton for arm and spine [23]. Origami robots are fabricated through folding processes, offering a straightforward yet versatile approach to develop diverse actuator morphologies. Inspired by biological systems, the integration of origami folding techniques with smart material actuators has demonstrated the capability to achieve dynamic expansion and contraction motions under pressurization and depressurization, respectively. The folding structures inherent to origami-based systems enhance their ability to conform to irregular surfaces, exhibiting properties characteristic of soft-bodied structures. For instance, researchers have developed origami-inspired soft pneumatic actuators capable of complex twisting motions for various applications, including medical devices [24]. Furthermore, origami and kirigami design methodologies employ top-down parallel transformation strategies, such as folding, to construct intricate and functional structures. These approaches draw inspiration from the natural environment, enabling the creation of complex geometries with enhanced adaptability and functionality. The combination of elegant design and mechanical versatility allows origami robots to perform sophisticated tasks while maintaining simplicity in their fabrication process. A notable example is the development of soft robotic deployable origami actuators for neurosurgical brain retraction, which offer precise control and adaptability during surgical procedures [25]. From an energy perspective, these actuators can be powered by compressed air, fluids, or magnetic fields, reducing reliance on complex electric motors and improving overall energy efficiency.

Based on the conducted research, we identified the design requirements that the device must support in order to be effective for the problem under consideration.

The device aims to fulfil the therapeutic function for pelvic floor rehabilitation and monitoring by proposing a system that facilitates remote therapy and the transmission of information and feedback between doctor and patient.

Through the use of soft actuators, a system has been developed that responds to pelvic floor muscle contractions and can react autonomously to the input received from the muscle, without the need for additional interfaces.

The softness and adaptability of the soft actuators, resulting from the choice of materials and the surface folding technique inspired by origami, enable the element to change significantly in size. In addition, it is necessary to position all the electronic components outside the vaginal canal to ensure both the safety of the user and the full effectiveness of the functions described above. The patient is also required to undergo physiotherapy concurrently, and the involvement of a psychologist is essential for comprehensive care. Many aspects of the project have been designed to integrate these two treatment approaches. Special attention will be given to the device's size, portability, and ease of use.

## 3 Results

### 3.1 *Carefit, Origami-Inspired Soft-Actuated Smart System*

The design process has led to the development of *Carefit*, a portable device for physiotherapeutic rehabilitation of the pelvic floor. Carefit can be subdivided into two relevant parts: a master unit composed of a recycled polypropylene shell where electromechanical components of the object's functioning reside, and a soft element made of biocompatible medical silicone, which functions as inflatable actuator (see Fig. 2).

The last one acts as an analogue vaginal probe and is inserted into the vagina before activating the device at the start of therapy. The inflatable actuator has a pattern of folds on its surface, taken from the art of origami, which allows the element in its initial condition to be very small and thus easy to insert (see Fig. 3).

Upon activation of the device, the patient will be required to perform contraction and relaxation exercises of the pelvic muscles, the inflatable actuator inserted into the vagina will thus receive input from the muscle contraction. With the presence of a pressure sensor, the system will be able to sense the pelvic muscle force exerted by the patient, process it into data and thus return a physical output. Upon muscular contraction, the actuator will react by inflating in proportion to the perceived pressure,

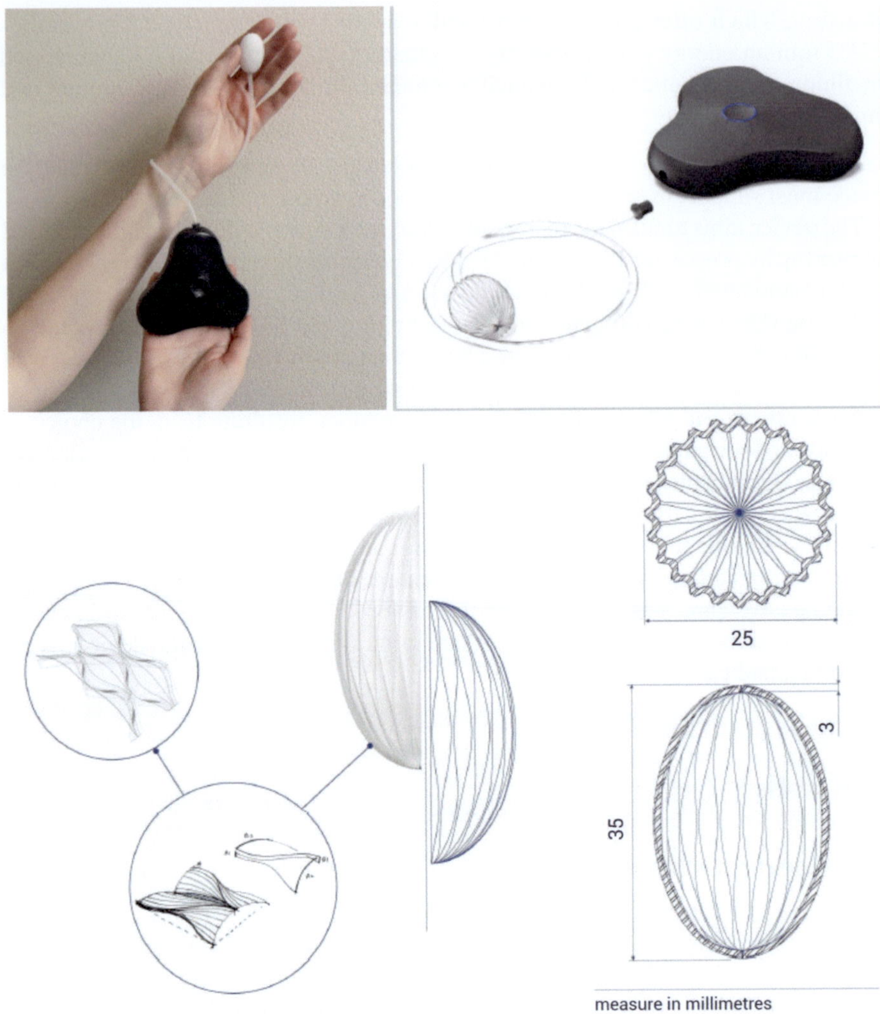


**Fig. 2** Carefit device: master unit and inflatable actuator

translated into a return airflow calculated by a pressure sensor, conversely, when the pelvic muscle is relaxed, the actuator will gradually return to its initial size.

The correlation between muscle force and inflatable actuator pressure is as follows: an increase in perceived muscle force will lead to a rise in pressure inside the actuator, causing it to inflate progressively. Conversely, a decrease in muscle strength will result in a reduction in pressure inside the actuator, causing it to deflate. The air injected into the actuator, passing through a temperature-controlled heating element will ensure the actuator element reaches the desired temperature to promote muscle relaxation. This type of treatment has great benefits for the patient, who will be able to independently control and set the limits of the training.

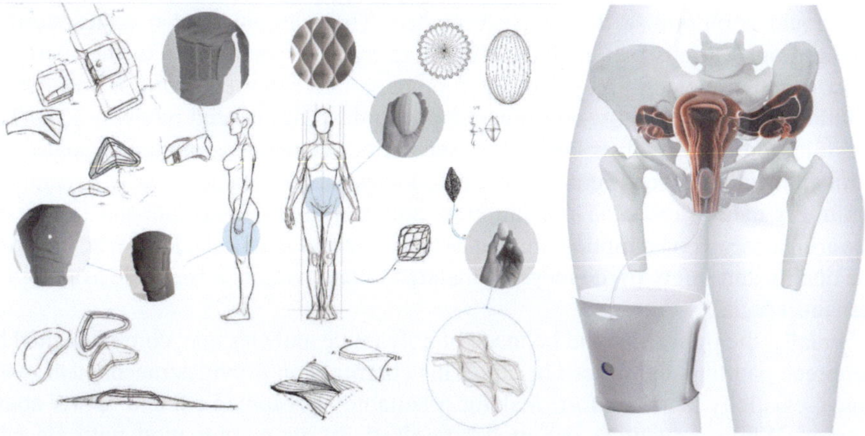

**Fig. 3** The origami-inspired inflatable actuator design

All input and output data will be recorded by the device and shared via an application with the patient and the treating physician.

Consequently, the device will be programmed according to the patient's needs, with parameters defined by a specialist after the visit.

The dimensions of the device are specifically designed to respect the user's body characteristics with the aim of designing a product that follows physical anatomical lines and does not create any psychological impediment in its wearability.

Some fundamental design aspects were considered to realise the project:

**Dimensions**. Particularly with regard to the balloon element, it was essential to start with very small dimensions (20 × 20 × 35 mm). Small dimensions can help the patient overcome a possible psychological block dictated by a compromised emotional and psychological state or by physical problems.

**Anatomical shape**. Being a wearable device, the design must respect wearability characteristics that include comfort, safety and speed of positioning. Regarding the positioning, the device must be placed on the thigh to facilitate the insertion of the balloon element into the affected area. A specially designed tight band secures the device in place. Additionally, the device's compact size allows the band to be worn discreetly under clothing.

**Materials**. Biocompatible silicone represents a key material in the medical and biotechnological sectors, so it is the most suitable material for the inflatable actuator. Due to its biocompatibility, it is well tolerated by the human body and is employed in devices such as prosthetics, surgical instruments, and catheters [26]. Its flexibility and high mechanical strength make it particularly suitable for wearable medical devices, while its chemical inertness prevents undesirable reactions with biological substances. Additionally, it ensures long-term biostability, preventing degradation that could compromise device performance. The material is also easily sterilizable through various methods, ensuring high hygiene standards. For the shell, the most suitable choice is recycled polypropylene that represents an eco-sustainable solution derived from post-consumer or post-industrial material recovery [27]. The recycling process involves collection, separation, cleaning, granulation, and reuse through mechanical or chemical methods. Its application reduces waste generation, optimizes natural resource consumption, and minimizes environmental impact by lowering energy consumption and the demand for raw materials. However, ensuring the quality and safety of the recycled material, along with its large-scale availability, remains crucial.

Finally, the tight strap will be made of polyamide and elastane, complemented by neoprene padding and Velcro inserts. This combination provides mechanical resistance, elasticity, and comfort, making it suitable for biomedical and sports applications. While polyamide has moderate elasticity, its combination with elastane enhances flexibility, improving adaptability to the user's morphology. Furthermore, it exhibits high moisture resistance and rapid drying properties, essential for sportswear and wearable devices [28].

The integration of these materials and manufacturing technologies enables the development of innovative and sustainable solutions for the Carefit project, ensuring safety, functionality, and efficiency in the biomedical and technological sectors.

**Smartphone application**. To support the device, the user can have access to the Carefit application to view detailed information regarding progress and the progress of training sessions. The Carefit application has many sections that educate and guide the user not only during exercise but also explain, through the display of illustrations and anatomical tables, in detail, the characteristics and consequences of each type of FGM. The application also acts as a means of user-doctor communication, the latter being able to view the patient's data and medical history and monitor the progress of the therapy.

## *3.2 Technological Components and Operation*

Carefit is a smart system composed of two main components: the connected digital electromechanical device and the telemedicine-oriented data management and sharing platform.

The electromechanical device serves both as a physical support for physiotherapy and as a tool for monitoring progress throughout a rehabilitation period. It features an active pneumatic master unit, where a micropump inflates an inflatable actuator while a solenoid valve deflates it based on data acquired from a pressure sensor that continuously monitors the patient's response.

The micropump is a specialized health-grade component that operates within a voltage range of 4–12 V with a flow rate of 0.4–1.0 l/min. During a therapy session, actuation is estimated to occur approximately every 10 s for 2 s to maintain the internal pressure at the therapeutic level. The inlet pipe is connected via a manifold to both the pressure sensor and the deflation control solenoid valve, which is also expected to operate with the same frequency and duration.

The air delivered to the inflatable actuator passes through a heating chamber that maintains the intake temperature between 37 and 38 °C. After a preheating phase of up to 60 s, the heating plate operates intermittently to ensure a stable temperature, with an estimated active time of approximately 20% of the total session duration.

The device is powered by a rechargeable 3.7 V lithium polymer battery, with step-up circuits providing the required 5–6 V voltage. The electromechanical system is controlled by a microprocessor unit (MPU) equipped with low-energy Bluetooth (BLE) connectivity and an RGB LED indicator that displays different operational statuses through flashing signals.

As previously said, the device is designed to respond instantly to the patient's physiological reactions, adjusting the pressure required by the rehabilitation program; during relaxation phases between contractions, the internal pressure decreases accordingly, causing the inflatable actuator to deflate to the minimum therapeutic pressure while ensuring a gradual increase over time to support muscle strengthening. All collected data is stored in integrated flash memory and synchronized via Bluetooth with a mobile app at the end of each session. This enables both the evaluation of therapy effectiveness and the secure sharing of data with the medical team, who can access the web platform through password-protected authentication.

The following table has been filled to estimate energy consumption and define the battery pack size, based on a hypothetical 30-min session (see Table 1).

Considering potential critical factors that may reduce the autonomy of LiPo cells—such as low temperatures, performance degradation, and occasional usage beyond the estimated average—it was decided to double the battery pack capacity compared to the estimation. Given that the typical usage scenario involves one session per day followed by a full device recharge, continuous use of the device for 30 min requires 231 mAh. By doubling this value, the required capacity becomes 462 mAh, so the best choice for this device can be a 500 mAh LiPo cell measuring 10 × 19 × 29 mm.

The LED indicator on the upper casing provides information about the device's status, serving as a guide for proper usage. A flashing green light indicates that the device is switched on and actively operating for the training session. A flashing blue light signals that the device needs to be paired via Bluetooth with a smartphone. A flashing red light alerts the user that the battery is low and requires recharging. Finally, a solid red light indicates a system malfunction requiring technical intervention.

## 4 Discussion

The aim of this work is to design an innovative medical device that combines advanced design principles and soft robotics for pelvic floor rehabilitation. The project focuses on developing a safe, adaptable, and sustainable therapeutic system

**Table 1** Estimate device energy consumption

| Component (operation) | Current I (mA) | Current V (V) | est. op. time (sec) | n. of op. | Consumption (mAh) |
|---|---|---|---|---|---|
| Micropump (initial inflate/ final deflate) | 250 | 6 | 120 | 1 | 14 |
| Micropump (pressure adjust op.) | 250 | 6 | 2 | 180 | 41 |
| Thermal plate (preheating) | 1400 | 5 | 30 | 1 | 16 |
| Thermal plate (temperature adjust op.) | 1400 | 5 | 360 | 1 | 189 |
| MPU (sensor management mode) | 150 | 5 | 1800 | 1 | 101 |
| MPU (BLE sync mode) | 300 | 5 | 180 | 1 | 20 |
| Solenoid valve (pressure adjust op.) | 450 | 5 | 2 | 180 | 61 |
| RGB led | 60 | 5 | 900 | 1 | 81 |
| TOTAL (mAh) | | | | | **462** |

capable of effectively addressing the clinical and functional needs of patients. From the comparative analysis of the current tools on the market (see Fig. 4), the project primarily considered the following key parameters:

**Responsive muscle contraction system**. The soft actuator is designed to autonomously respond to input generated by pelvic floor muscle contractions, eliminating the need for external interfaces. This ensures immediate and personalized feedback during therapy sessions.

**Integrated data exchange system**. The system includes a digital platform for data collection, processing, and transmission (smartphone application). This system enables data sharing between the patient and the medical staff, optimizing treatment monitoring and customization (see Fig. 5).

**Dimensional optimization**. The soft actuator features compact initial dimensions (35 mm in height and 20 mm in width) to facilitate insertion and ensure anatomical adaptability. Surface folds allow the actuator to expand up to 95 mm in height and 55

PROS ✓
CONS ✗

| | VAGINAL PROBE | VAGINAL DILATORS | SEX TOYS | CAREFIT |
|---|---|---|---|---|
| ANATOMICAL ADAPTABILITY | ✗ | ✗ | ✗ | ✓ |
| INTERACTIVE | ✗ | ✗ | ✓ | ✓ |
| MYO-RESPONSIVE | ✗ | ✗ | ✗ | ✓ |
| PORTABILITY | ✓ | ✓ | ✓ | ✓ |
| ACCEPTABILITY | ✓ | ✓ | ✗ | ✓ |
| DISCRETION | ✓ | ✓ | ✗ | ✓ |
| PERCEIVED NOISE | ✓ | ✓ | ✗ | ✗ |
| CUSTOMISABLE | ✗ | ✗ | ✗ | ✓ |
| TELEMEDICINE INTEGRATION | ✗ | ✗ | ✗ | ✓ |
| DATA LOGGING/ VISUALIZATION | ✗ | ✗ | ✗ | ✓ |
| EASE OF MAINTENANCE | ✓ | ✓ | ✗ | ✗ |

**Fig. 4** Comparison with existing devices

mm in width, effectively responding to compression and decompression movements during inflation.

**Safety-focused design**. All design choices prioritize patient safety. The device's electromechanics components are separated from the soft element inserted into the vaginal canal, eliminating direct contact with inner organs. Additionally, the device's shape has been designed to minimize any potential risks during use.

**Environmental, social, and economic sustainability**. The project adopts a sustainable approach at every stage, from material selection to the device's end of life. The actuator is manufactured from biocompatible medical-grade silicone, while initial prototypes were developed using 3D-printed TPU. The protective casing, which houses the electromechanical components, is made of recycled polypropylene to reduce environmental impact.

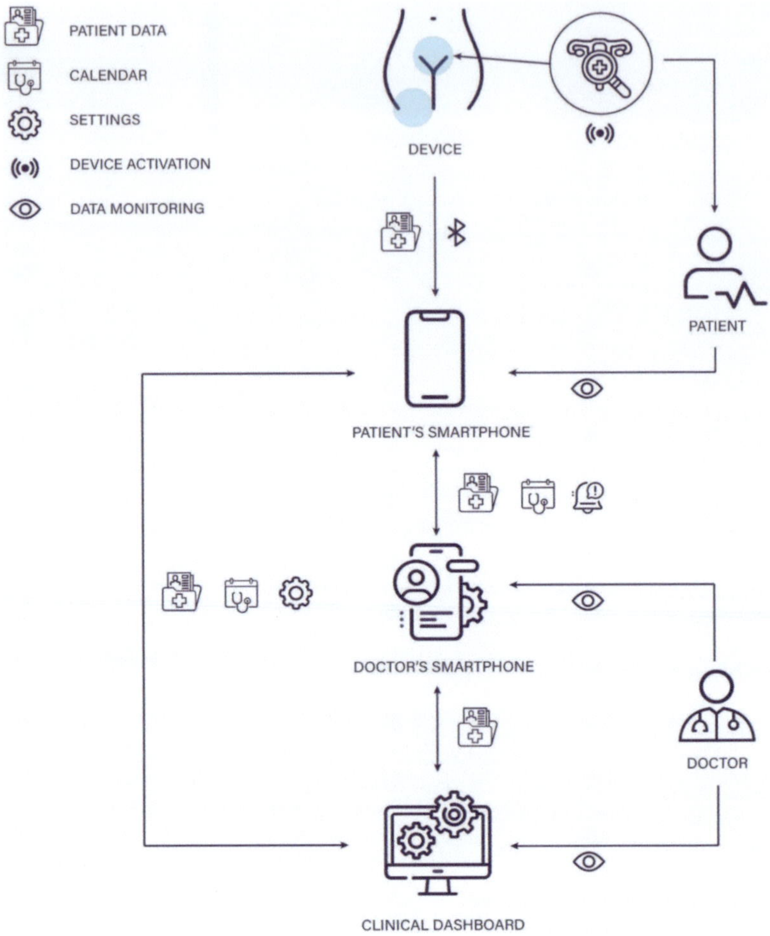


**Fig. 5** Carefit system architecture

The inflatable actuator, designed for insertion into the vaginal canal, features an elliptical cross-section to facilitate insertion and optimize inflation performance. Its compact starting dimensions ensure ease of use, while its expandable design provides adaptability to individual anatomical variations. The electronic components are housed in a curved casing designed to fit the patient's thigh area, secured within an adjustable band (see Fig. 3). This configuration enhances mobility during physiotherapy sessions, simplifies wearability, and minimizes perceived bulk.

In summary, this device represents a technologically advanced and patient-centric solution for pelvic floor dysfunction treatment. Its design ensures safety, adaptability, and sustainability while providing an optimized therapeutic experience and addressing critical clinical requirements. This innovative approach has the potential to significantly improve outcomes in pelvic floor rehabilitation while aligning with contemporary environmental and healthcare priorities.

The limitations of this work mainly relate to clinical trials and usability validation. To date, clinical trials have not been conducted to assess its therapeutic efficacy, safety and adaptability. Thus, future research will prioritize these clinical evaluations. Additionally, beyond clinical validation, further studies will focus on usability aspects to ensure that the system is not only functionally and clinically effective but also user-friendly, intuitive and culturally acceptable. Furthermore, all the implications related to the approval and certification process that medical devices must undergo should not be underestimated.

## 5 Conclusion

Advancements in telemedicine for pelvic floor rehabilitation mark a significant step toward more personalized, accessible, and effective therapeutic interventions. The integration of innovative systems such as Carefit, which combines electromechanical components with digital platforms for data management, presents new opportunities for real-time monitoring and dynamic therapy optimization. By leveraging dedicated mobile applications, patients can receive continuous feedback, enabling healthcare providers to adjust treatments based on individual responses, thereby enhancing clinical outcomes and promoting a patient-centered approach.

The use of soft actuators, inspired by advanced design methodologies, such as origami, offers greater levels of safety, precision, and adaptability, ensuring effective and non-invasive interaction with the human body. These technological advancements not only improve therapeutic efficacy but also address cultural sensitivities, fostering inclusivity and acceptance across diverse patient populations.

Additionally, the broader application of telemedicine in pelvic floor rehabilitation holds the potential to overcome logistical and economic barriers, ensuring access to high-quality care in remote and underserved areas. The adoption of biocompatible materials, sustainable design principles, and scalable technologies further enhances the impact of these innovations. Such developments pave the way for addressing a wide spectrum of pelvic floor dysfunctions, ultimately improving the quality of life and overall well-being of patients while aligning with contemporary priorities in healthcare and sustainability.

## References

1. World Health Organisation (WHO): Female Genital Mutilation. https://www.who.int/en/news-room/fact-sheets/detail/female-genital-mutilation. Last accessed 27 Jan 2025

2. Ghani, A., Abi Ahmed, L., Busoni, M., Consoli, D., Ndjaye, D., Sacchetti, G., Trugli, C.: Profilo Informativo del Fenomeno della Mutilazioni Genitali Femminili, Conoscerle per Prevenirle. Linee guida Regione Toscana (2006)
3. Abdulcadir, J., Sachs Guedj, N., Yaron, M.: Female Genital Mutilation/Cutting in Children and Adolescents: Illustrated Guide to Diagnose, Assess, Inform and Report [Internet]. Springer, Cham (CH) (2022)
4. UNFPA-UNICEF, Joint Program on the Elimination of Female Genital Mutilation Annual Report 2019. https://www.unfpa.org/publications. Last accessed 30 Jan 2025
5. Powers, S.A., Burleson, L.K., Hannan, J.L.: Managing Female Pelvic Floor Disorders: A Medical Device Review and Appraisal. Interface Focus (2019)
6. Liu, M., Juravic, M., Mazza, G., Krychman, M.L.: Vaginal Dilators: Issues and Answers, vol. 9, Issue 2, pp. 212–220. Sexual Medicine Review (2021)
7. Chen, P.H., Li, Y.M., Morris, K., Makale, M.T., Mayadev, J., Talke, F.E.: Design and Material Characterization of an Inflatable Vaginal Dilator. Materials (Basel) (2024)
8. Cullen, K., Fergus, K., DasGupta, T., Fitch, M., Doyle, C., Adams, L.: From "sex toy" to intrusive imposition: a qualitative examination of women's experiences with vaginal dilator use following treatment for gynecological cancer. J. Sex. Med. **9**(4), 1162–1173 (2012)
9. Xu, D., Zhang, M., Xu, H., Fu, J., Li, X., Xie, S.Q.: Interactive compliance control of a wrist rehabilitation device (WReD) with enhanced training safety. J. Healthc. Eng. 1–10 (2019)
10. Pan, M., Yuan, C., Liang, X., Dong, T., Liu, T., Zhang, J., Jun, Z., Huayong, Y., Bowen, C.: Soft actuators and robotic devices for rehabilitation and assistance. Adv. Intell. Syst. **4**(4), (2022)
11. Peng, Z., Huang, J.: Soft rehabilitation and nursing-care robots: a review and future outlook. Appl. Sci. **9**(15), (2019)
12. Huang, J., Liu, N., Wang, H., Cui, L., Bai, N., Tian, S.: Design and simulation of spine rehabilitation soft robotic actuator. In: 5th International Conference on Control, Automation and Robotics (ICCAR), pp. 188–193. Beijing (2019)
13. Romero, A.M.L., Valbuena, Y., Velasco, A., Solaque, L.: Soft-actuated modular knee-rehabilitation device: proof of concept. In: Proceedings of the 4th International Conference on Bioinformatics Research and Applications, pp. 71–78 (2017)
14. Dragone, D., Randazzini, L., Stano, G., Capace, A., Nesci, F., Cosentino, C., Merola, A., et al.: Design, testing and control of a smart haptic interface driven by 3D-printed soft pneumatic actuators for virtual reality-based hand rehabilitation. Smart Mater. Struct. **32**(4), 045009 (2023)
15. Quan, L., Jie, Z., Chang, Z., Sheng, Q.X.: Design and control of soft rehabilitation robots actuated by pneumatic muscles: state of the art. Fut. Gener. Comput. Syst. **113**, 620–634 (2020)
16. Shin, S.Y., Hohl, K., Giffhorn, M., Awad, L.N., Walsh, C.J., Jayaraman, A.: Soft robotic exosuit augmented high intensity gait training on stroke survivors: a pilot study. J. Neuroeng. Rehabil. **19**(1), 51 (2022)
17. Hu, X., Zeng, X., Xu, Y., Luo, C., Jia, L., Zhao, Z., Sun, Z., Qu, X.: A soft robotic intervention for gait enhancement in older adults. IEEE Trans. Neural Syst. Rehabil. Eng. **29**, 1838–1847 (2021)
18. Buffagni, A.: Designing a soft-actuated smart garment for postural control and fall prevention in elderly women. In: EAI International Conference on IoT Technologies for HealthCare, pp. 121–135. Springer Nature Switzerland, Cham (2022)
19. De Pascali, C., Naselli, G.A., Palagi, S., Scharff, R.B., Mazzolai, B.: 3D-printed biomimetic artificial muscles using soft actuators that contract and elongate. Sci. Robot. **7**(68) (2022)
20. Commissione Europea: Muscoli artificiali stampati in 3D in un unico passaggio.CORDIS. https://cordis.europa.eu/article/id/442717-3d-printing-artificial-muscles-in-a-single-step/it. Last accessed 29 Jan 2025
21. Chauhan, M., Chandler, J.H., Jha, A., Subramaniam, V., Obstein, K.L., Valdastri, P.: An origami-based soft robotic actuator for upper gastrointestinal endoscopic applications. Front. Robot (2021)
22. Liu, S., Fang, Z., Liu, J., Tang, K., Luo, .J, Yi, J., Hu, X., Wang, Z.: A compact soft robotic wrist brace with origami actuators. Front Robot AI (2021)

23. George, D., Kawamura, S., Zheng, Y., Hayashi, Y.: Design and validation of a high DOF origami inspired soft actuated assistive exoskeleton arm and spine. IEEE Trans. Med. Robot. Bion. **5**, 1045–1056 (2023)
24. Li, D., Fan, D., Zhu, R., Lei, Q., Liao, Y., Yang, X., Pan, Y., Wang, Z., Wu, Y., Liu, S., Wang, H.: Origami-inspired soft twisting actuator. Soft Rob. **10**(2), 395–409 (2023)
25. Amadeo, T., Van Lewen, D., Janke, T., Ranzani, T., Devaiah, A., Upadhyay, U., Russo, S.: Soft robotic deployable origami actuators for neurosurgical brain retraction. Front. Robot. AI **8**, 731010 (2022)
26. Ratner, B.D., Hoffman, A.S., Schoen, F.J., Lemons, J.E: Biomaterials Science: An Introduction to Materials in Medicine. Elsevier (2013)
27. Seyednejad, H., Ghassemi, A.H., van Nostrum, C.F., Vermonden, T., Hennink, W.E.: Functional aliphatic polyesters for biomedical and pharmaceutical applications. J. Control. Release **152**(1), 168–176 (2011)
28. Drobny, J.G.: Handbook of Thermoplastic Elastomers. Elsevier (2014)

# Top-Forging Crack Detection for Cold-Heading Steel

**Long Wu, Yu-Xiu Wu, and Hong Wei Yang**

**Abstract** Cold heading steel top forging cracks are challenging to capture due to their complex shapes and varying structural dimensions. Additionally, differences in understanding standards and reference images can lead to inconsistent judgment results in manual inspection. This paper introduces a detection and classification algorithm for cold-heading steel top forging cracks based on an enhanced YOLOv8n model, incorporating multiple attention mechanisms to address these challenges. The proposed algorithm first integrates receptive field attention convolution (RFAConv) and CBAM (concentration-based attention module) into the backbone network to improve the convolutional modules. Moreover, DAttention is added to the backbone network for dynamically allocating attention weights and speeding up model training. In the neck portion of the network, C2f and Conv modules are replaced with improved C2f_RFCBAM and RFCBAMConv modules to optimize the model and further increase detection precision. Finally, a new loss function, SioU, is introduced to accelerate convergence and improve regression accuracy. Experimental results show that the improved YOLOv8n model applied to a self-annotated cold heading steel dataset, achieves increases of 7.3% in mAP@0.5, 2.3% in mAP@0.5:0.95, and 10.3% in recall rate.

**Keywords** Cold heading steel · Surface defect detection · YOLOV8n · Sensory field attention · Deformable attention mechanism

## 1 Introduction

Cold-heading steel is a steel material used to process fasteners. Fasteners are essential spare parts for fastening connections that are widely used and are figuratively called the "rice of industry" [1]. With the increasing variety of cold-heading steel and the

L. Wu · Y.-X. Wu (✉) · H. W. Yang
School of Electrical and Information Engineering, Anhui University of Technology, Ma'anshan, China
e-mail: Yuxiuwu@ahut.edu.cn



Y. Ma (ed.), *Proceedings of 4th International Conference on Intelligent Systems Design and Engineering Applications*, Lecture Notes in Electrical Engineering 1478,
https://doi.org/10.1007/978-981-95-1802-9_12

upgrading of product grade, the requirement for product quality is also increasing. How to detect the quality quickly and accurately has become a problem that must be faced at this stage. Top forging inspection is one of the most important means of cold-heading steel quality inspection in the industry nowadays, as it can quickly and accurately inspect the product quality by simulating the deformation of the user in the process of use. Currently, the judging methods of top forging in each steel mill are mainly based on manual visual judgment. In addition, the side shape of the top forging samples is mainly based on cracking, and the content of the shape is judged according to the opening angle, size, depth, and other characteristics of the cracks.

Existing defect detection methods, including eddy current, infrared, and magnetic leakage, all exhibit certain limitations when detecting and classifying defects in wire top forging. Although the eddy current detection technique can recognize defects in top forging, it cannot further screen these cracking patterns, limiting its application scope [2]. Infrared detection technology has high requirements for the stability of the detection environment and the configuration of the computer system, which may bring inconvenience and challenges in practical application [3]. Magnetic leakage detection technology, unable to detect small or narrow crack defects, makes it challenging to meet the demand for high-precision detection of acceptable defects in the top forging [4]. In summary, the limitations of traditional detection mean that it cannot meet the strict requirements of crack detection in the top forging of wires.

In recent years, thanks to the development of computer technology, data-supported machine-learning algorithms have flourished, and the application of machine learning in the iron and steel industry has become increasingly widespread [5–8]. The method can automatically extract features from images and has performed well in tasks such as classification and localization [9]. To address the low efficiency and accuracy of current surface defect detection methods, Neng [10] proposed a model system based on improved YOLOv5s performing well in visual target detection. Tabernik et al. [11] proposed a segmentation-based deep learning framework for detecting and segmenting surface anomalies; the framework's design allows the model to be trained using a small number of samples for training, which is crucial for real-world application scenarios. Liu et al. [12] proposed an advanced target detection model that introduces a feature pyramid network with a multi-scale design to efficiently capture various targets regardless of their size to improve their adaptability. To address the problem of information about the candidate region being lost before applying the classifier, Chaudhari et al. [13] compensated for the loss of important feature information in the candidate region before the classifier. This method has more satisfactory results for detecting cracks on the surface of considerable size, size-changing objects, and small objects. Xu et al. [14] optimized the C3 module in the YOLOv5s network by using the CBAM attention mechanism and then by using CARAFE as an upsampling module, and at the same time, proposed a pyramid structure incorporating a cross-stage localized network to improve the detection accuracy of the network.

Given the challenges faced in cold heading steel target inspection tasks between the complexity of the dataset and the limited hardware resources, it is still challenging to maintain high performance while considering the rational utilization of resources despite the improvement of existing methods. In this paper, a cold heading steel crack dataset was collected and produced according to the field test environment. On this basis, the YOLOv8n [15] model from YOLOv8 was chosen as the foundation, an optimization model integrating multiple advanced attention mechanisms is proposed, which can significantly improve the model's detection of cracks by optimizing the improved convolutional module, the new attention mechanism module, and the loss function in the backbone network. First, RFCBAM (Refined Convolutional Block Attention Module) is optimized by Receptive Field Attention Convolution.

RFAConv (Receptive et al.) [16] is introduced to replace the original backbone network and the standard convolutional layer in the neck. At the same time, in order to further strengthen the effectiveness of feature extraction, the backbone network and the C2f module in the neck are reconstructed and upgraded. By adjusting the internal structure of the module and introducing a more efficient feature fusion and transformation mechanism, the C2f module can output richer and more discriminative feature maps while maintaining its lightweight. Second, to further improve the accuracy and efficiency of crack detection in cold heading steel, DAttention (deformable attention mechanism) is incorporated into the backbone network [17]. Finally, to improve the crack detection model's convergence speed and regression accuracy in cold heading steel, a new loss function, SioU (Simplified Intersection over Union), is introduced [18]. The SioU loss function performs well in target detection and measures the difference between the predicted and actual frames by integrating multiple geometrical properties between the bounding boxes. The difference between the predicted and actual frames is measured by combining multiple geometric properties between the bounding boxes.

## 2 YOLOv8n Improved Model

For the problem of a few pixels occupied by cold heading steel crack targets and the imbalance of categories, YOLOv8n suffers from low accuracy and leakage detection. For the existing problems and the characteristics of this dataset, the improvement of this base model mainly includes:

(1) Use RFAConv to change the original CBAM to get a brand new RFCBAM module to replace the convolutional part in the backbone and neck, as well as designing a brand new C2_RFCBAM to replace the C2f module in the backbone and neck, which enhances the feature extraction ability of the backbone network and the fusion ability of the neck.
(2) The DAttention attention mechanism is introduced into the backbone network to dynamically assign the attention weights, which accelerates the model training process and enables the model to focus more on the crack information in the image, thus further improving the detection accuracy.
(3) Under the premise of introducing the DAttention attention mechanism in the backbone network to allocate the attention weights dynamically and combined with the use of the new loss function SIoU (Scalable Intersection over Union

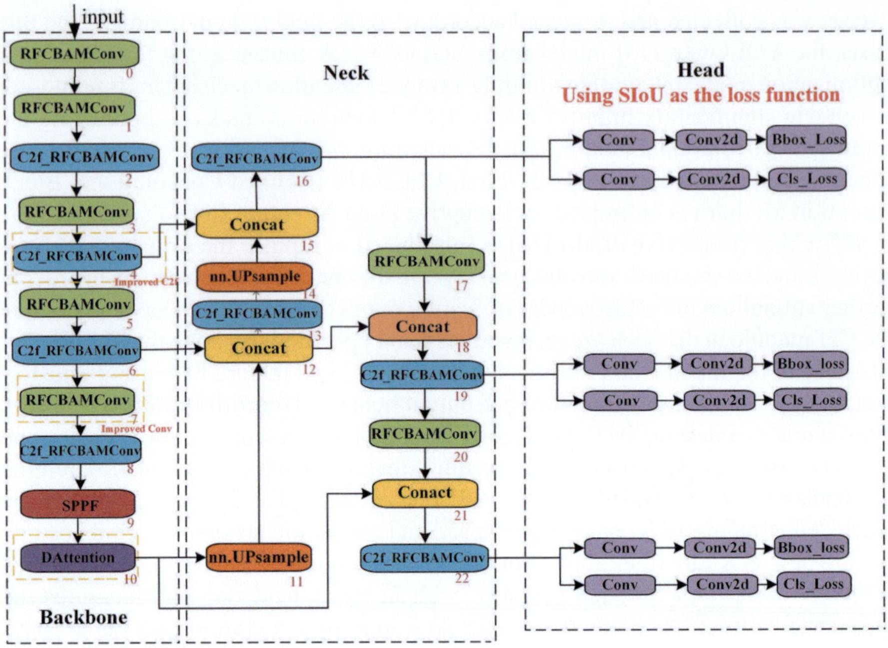


**Fig. 1** Improved YOLOv8n model structure

Loss), the convergence speed and regression accuracy of the model can be significantly improved, which in turn enhances the accuracy of the detection of the crack information in the image. The improved model network structure is shown in Fig. 1.

## *2.1 Optimization of CBAM and C2f Modules*

CBAM has the shortcomings of computational redundancy, so it is improved, and for the sake of narrative convenience, the improved CBAM is called RFCBAM. CBAM enhances the feature maps by sequentially applying the channel and spatial attention modules. First, the input feature maps are compressed by global average pooling and global maximum pooling to obtain channel attention weights; then, the original feature maps are weighted using these weights, followed by applying the spatial attention module to refine the features further. RFCBAM, on the other hand, employs a more compact approach by weighting channel attention and spatial attention (in particular, the sensory field spatial features). It uses the SE (Squeeze-and-Excitation) attention mechanism to replace part of the channel attention component in BAM. It introduces grouped convolution to extract the sensory field spatial features to reduce computational overhead and feature overlap. The CBAM module and RFCBAM module pairs are shown in (a) and (b) of Fig. 2.

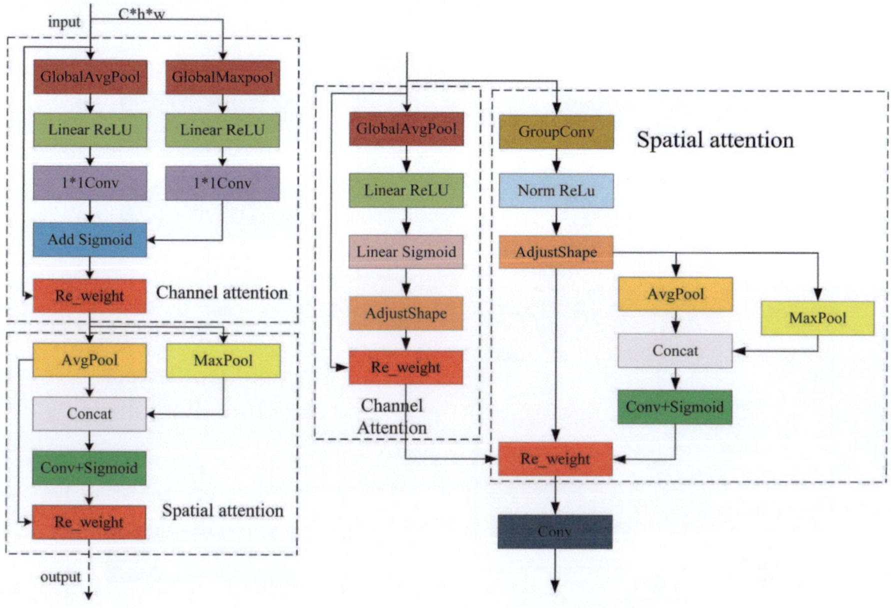


**Fig. 2** Comparison chart of CBAM and RFCBAM modules

By introducing the RFCBAM adjustment and attention mechanism to transform the Bottleneck structure in the original C2f module, the feature extraction capability of the model can be significantly improved. The specific structure is shown in Fig. 3. Specifically, the modified C2f module is named C2f_RFCBAM, and its core lies in using RFCBAM_neck to replace the traditional Bottleneck.

## 2.2 *DAttention Module*

The operation principle of the Deformable Attention mechanism (DAttention) and its revolutionization of the traditional Transformer model are shown in Fig. 4. The mechanism subverts the fixed and rigid allocation of attention in the traditional Transformer and instead adopts a more flexible and data-driven approach to adjust the attention weights dynamically.

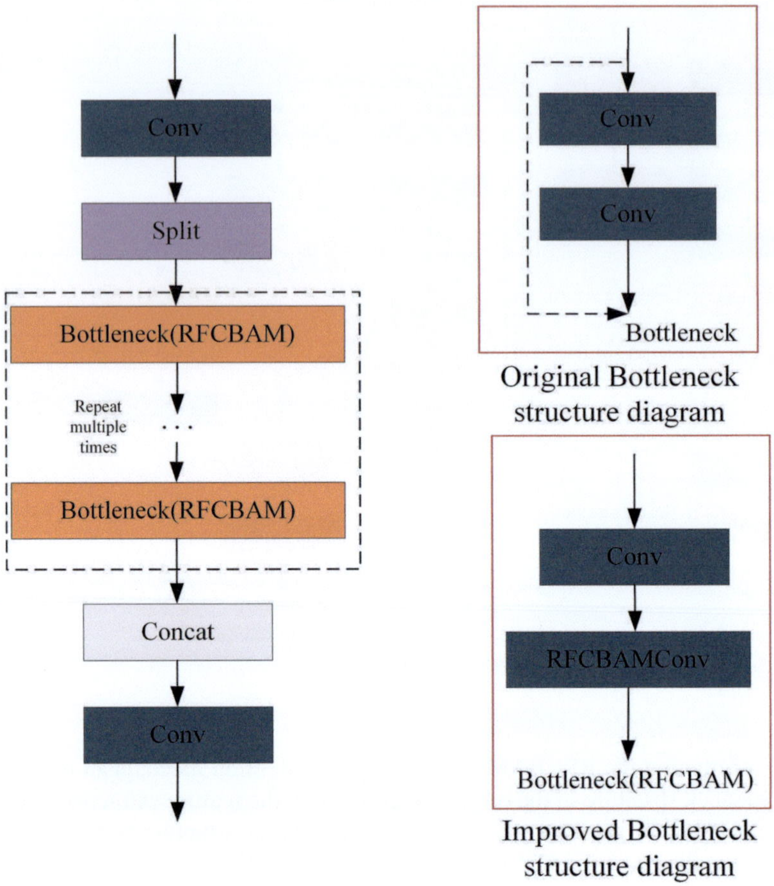


**Fig. 3** Schematic diagram of C2f_RFCBAM module and inner module

## *2.3 Loss Function Optimization*

Traditional loss functions, such as GIoU (Generalized Intersection over Union), CIoU (Complete Intersection over Union), etc., despite their contribution to improving detection performance, often face the problems of slow model convergence and instability of prediction frames during training (i.e., "drift"). To overcome these challenges, Gevorgyan et al. proposed an innovative SIoU loss function that combines multiple dimensions to optimize the match between predicted and real frames.

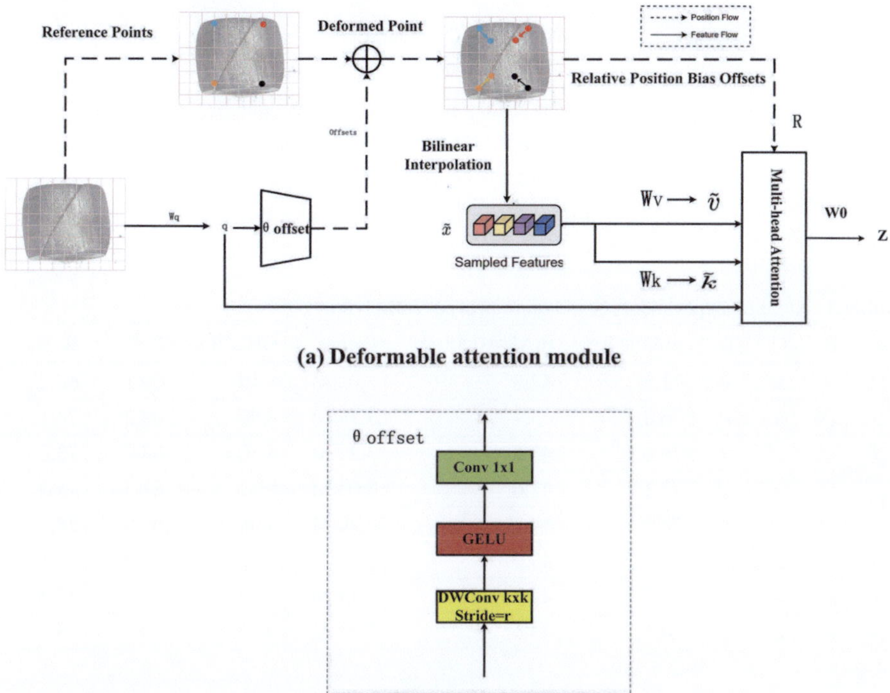


**(a) Deformable attention module**

**(b) Offset network**

**Fig. 4** DAttention structure diagram

# 3 Experimentation and Analysis

## *3.1 Performance Indicators and Ablation Experiments*

In order to realize cold-heading crack detection, this paper uses 3000 high-definition images of cold-heading steel cracks captured by Daheng's industrial camera. The camera captures the dataset in different light, angles, and complex scenes. The dataset contains 3000 images and is categorized according to the specifications into four categories: Flawless, 45-degree crack, micro-crack, and fragment. Moreover, the collected photos are divided into a training set (2400 photos), a validation set (300 photos), and a test set (300 photos) according to 8:1:1. The experimental training was performed on Windows 10, CUDA 12.5, Pytorch deep learning framework, and NVIDIA GeForce RTX 4070 Ti GPU.

Keeping the experimental environment unchanged, all models had an input image size of 640 $\times$ 640 pixels, and all underwent the same training, including using the same training parameters and cycles (200 epochs in total). For all the following training processes, all models were optimized and configured according to the

**Table 1** Training parameter settings

| Parameters | Configuration | Parameters | Configuration |
|---|---|---|---|
| epochs | 200 | Momentum | 0.934 |
| Batch | 16 | Weight_decay | 0.0005 |
| imgsize | 640 | Close_mosaic | 10 |
| Lrf | 0.01 | Workers | 8 |
| Lr0 | 0.01 | Workspace | 4 |

**Table 2** Ablation experiments based on improved YOLOv8n

| A | B | C | D | mAP@0.5 | mAP@0.5:0.95 | Param | GFLOPs | P (%) | R (%) |
|---|---|---|---|---|---|---|---|---|---|
| √ | × | × | × | 71.8 | 42.3 | 3.01 M | 8.10 | 69.1 | 68.2 |
| √ | √ | × | × | 76.8 | 43.4 | 3.14 M | 8.80 | 69.2 | 74.1 |
| √ | × | √ | × | 76.6 | 43.6 | 3.27 M | 8.30 | 67.9 | 73.8 |
| √ | × | × | √ | 73.1 | 42.6 | 3.03 M | 8.12 | 67.5 | 75.5 |
| √ | √ | √ | × | 78.7 | 44.3 | 3.40 M | 9.08 | 69.8 | 78.3 |
| √ | × | √ | √ | 76.1 | 43.9 | 3.28 M | 8.32 | 69.2 | 77.1 |
| √ | √ | × | √ | 77.4 | 44.1 | 3.14 M | 8.82 | 69.5 | 77.3 |
| √ | √ | √ | √ | 79.1 | 44.6 | 3.41 M | 9.10 | 70 | 78.5 |

settings listed in Table 1. Improvements were made in the YOLOv8n model, and ablation experiments were carried out on a homemade dataset. The results of each improvement were statistically analyzed, and the results are shown in Table 2.

In Table 2, A denotes the base model YOLOv8n, and B, C, and D denote the addition of RFCBAM, DAttention, and SioU to each of the YOLOv8n models, respectively. From the improved experimental results, adding B, C, and D effectively improves the indexes of Map @ 0.5.

## 3.2 *Algorithm Comparison Test*

In order to highlight the improvement effect of the model, the successive generations of YOLO models and the improved model are tested on the same experimental platform, and the effect comparison is shown in Fig. 5. In order to verify the detection ability of the improved YOLOv8 model, the improved model is compared with representative networks such as Faster R-CNN, YOLOv3-tiny, and YOLOv7-tiny. The experimental results are shown in Table 3. According to the experimental results, although Faster R-CNN is a classical two-stage target detection algorithm, the value of its mAP is 63.4% in the cold heading steel crack detection process, indicating that its ability to detect small targets is relatively low. P and R show low results, so they are not suitable for the application scenario of small defect detection. YOLOv3-tiny mAP values, P and R, are slightly lower than the Faster R-CNN detection algorithm, so they are also unsuitable for this dataset. Similarly, it is easy to see that other models have similar problems.

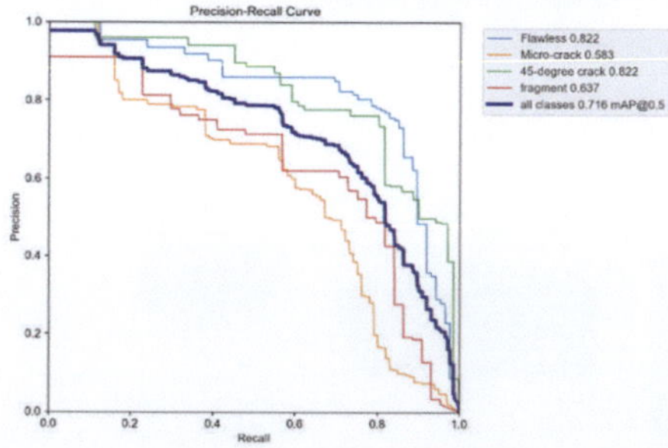


(a) YOLOv5n

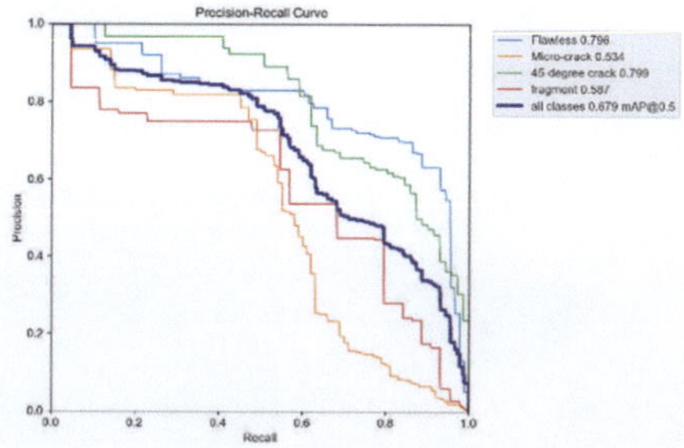


(b) YOLOv7-tiny

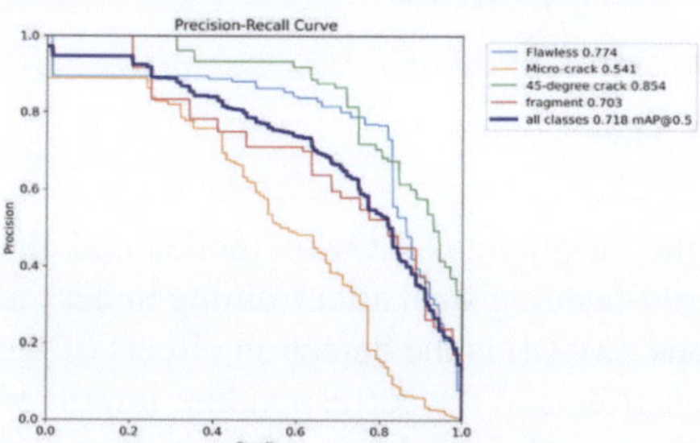


(c) Basic YOLOv8

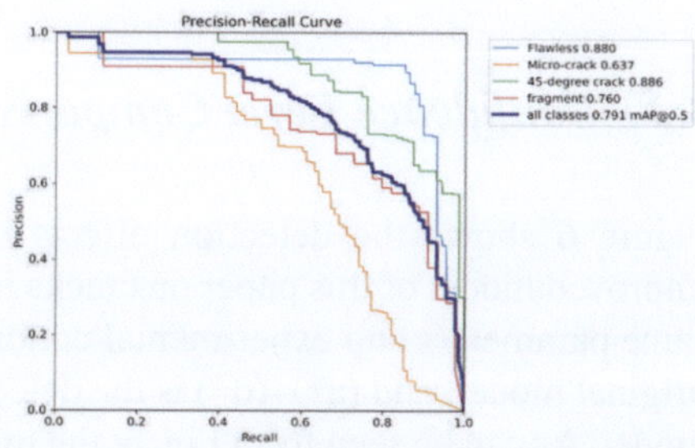


(d) Improvement of YOLOv8 model in this paper

**Fig. 5** PR curves of each model

**Table 3** Algorithm comparison experiment

| Models | Precision (%) | Recall (%) | GFLOPs | Param | mAP@0.5 | FPS |
|---|---|---|---|---|---|---|
| Faster R-CNN | 59.3 | 62 | 41.38 | 41.38 M | 63.4 | 63 |
| RTDETR-L | 69.7 | 73.2 | 110 | 32.82 M | 75.3 | 89 |
| YOLOv3-tiny | 58.7 | 61.5 | 19.05 | 12.14 M | 62.1 | 134 |
| YOLOv5n | 68.2 | 67.3 | 7.1 | 2.5 M | 71.6 | 163 |
| YOLOv6 | 66.3 | 65.4 | 8.7 | 4.23 M | 68.5 | 145 |
| YOLOv7-tiny | 68.4 | 63.0 | 105.2 | 37.22 M | 67.9 | 138 |
| YOLOv8n | 69.1 | 68.2 | 8.10 | 3.01 M | 71.8 | 209 |
| OURS | 70 | 78.5 | 9.10 | 3.41 M | 79.1 | 151 |

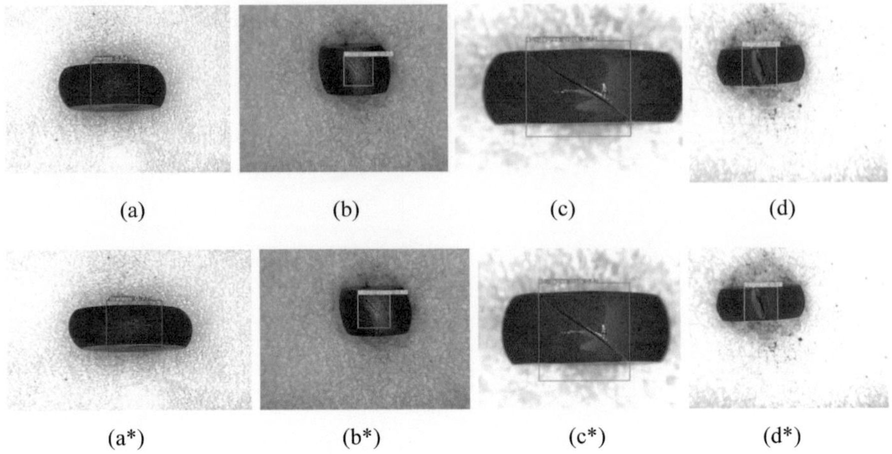


**Fig. 6** Comparison of test results before and after model improvement

## *3.3 Confidence Level Comparison Test*

Figure 6 shows the detection effects of the original YOLOv8n model and the improved model of this paper on cracks in cold-heading steel after training under the same parameters and experimental conditions. (a)–(d) is the detection effects of the original model, and (a*)–(d*) is the detection effects of the corresponding improved model. As can be seen from Fig. 6, the improved model can detect the corresponding crack information more accurately in the cold heading steel crack detection process. For example, the confidence of Flawless detection increases from 0.54 to 0.77, Micro-crack detection increases from 0.25 to 0.6, 45-degree crack detection increases from 0.73 to 0.87, and fragment detection increases from 0.59 to 0.79. This improved version is more accurate in detecting cracks. 0.59–0.79, this improved version has realized a leap forward in detection accuracy.

# 4 Conclusion

(1) This paper introduces an optimized YOLOv8n model featuring multi-attention fusion for crack detection in cold heading steel. By integrating receptive field attention convolution and the CBAM (concentration-based attention module) mechanism, the model enhances the convolutional module within the backbone network. It ensures consistency in receptive field features while addressing the challenge of shared attention weight parameters. Moreover, the model incorporates attention mechanisms in both channel and spatial dimensions of the features. Additionally, the DAttention mechanism is embedded within the

backbone network to dynamically allocate attention weights, thereby accelerating model training. Other innovative measures include improvements to the neck and loss function, significantly boosting the model's accuracy and overall capability in crack detection.

(2) This paper proposes a YOLOv8n optimization model incorporating multi-attention for cold-heading steel crack detection. The experimental results show that the optimized model improves Precision by 10.7, 11.3, 1.6, and 0.9% compared to representative networks such as Faster R-CNN, YOLOv3-tiny, YOLOv7n, and YOLOv8n on the cold heading steel cracking dataset, respectively. Recall improved by 16.5, 17, 15.5, and 10.3%, mAP@0.5 15.7, 17, 11.2, and 7.3%, respectively. This shows the advantages of the optimized model over the traditional representative network on the cold heading steel cracking dataset.

(3) In cold heading steel top forging inspection, this algorithm can detect the surface features of existing cracks, saving labor intensity. However, it still requires manual assistance for identifying fine cracks. By subsequently using sensors like eddy current to measure crack depth, full automation of crack detection in cold heading steel top forging can be achieved, enabling wide on-site application. This will also be the focus of our next steps.

## References


1. Liu, C., Nong, Z.J., Zhang, Z.M.: Analysis of cold top forging cracking problem of cold heading steel wire rod and improvement measures. Southern Metal **6**(207), 30–33 (2015)
2. Chen, W., Jiang, K., Zhang, Z., et al.: Application of eddy current testing technology in metal surface defect detection. World Nonferrous Metals **599**(11), 160–162 (2022)
3. Hao, W., Jia, L., Yazhong, S., et al.: Research and application progress of active infrared thermography in aviation engine blade defect detection. Laser Infrared **51**(12), 1554–1562 (2021)
4. Zhang, H., Li, J., Xie, F. et al.: Application of magnetic flux leakage detection technology in ferrite defect detection. Labor. Res. Explor. **41**(1): 57–61 (2022)
5. Liu, J.: Artificial intelligence drives changes in the metallurgical industry. Steel **55**(06), 1–7 (2020)
6. Yang, J., Wu, S.W.: Performance prediction of steel rolling process based on machine learning. Iron Steel **56**(09), 48–51 (2021)
7. Li, J.Y., Yang, Z.F., Zheng, J.F., et al.: Application of deep learning technology in the steel industry. Iron Steel **56**(09), 43–49 (2021)
8. Daizj, H.Y., Liu, W., et al.: Mask-RCNN-based topography detection of converter hearth. Iron Steel **58**(03), 73–78 (2023)
9. Lu, J.Z., Zhang, C.Y., Liu, S.P., et al.: Lightweight DCN Yolo for strip surface defect detection in complex environments. Comput. Eng. Appl. **59**(15), 318–328 (2023)
10. Deng, N.H., Zhou, B.G., Zhang, Z.J., et al.: A billet surface defect detection system based on the improved YOLOv5s. Instrument. Technol. Sens. **10**, 72–78 (2023)
11. Tabernik, D., Šela, S., Skvarč, J., et al.: Segmentation-based deep-learning approach for surface-defect detection. J. Intell. Manuf. **31**(3), 759–776 (2020)
12. Liu, W., Anguelov, D., Erhan, D., et al.: SSD: single shot multibox detector. In: Computer vision–ECCV 2016: 14th European Conference, Amsterdam, The Netherlands, October 11–14, 2016, Proceedings, Part I 14. Springer International Publishing, pp 21–37 (2016)

13. Chaudhari, C.V.: Steel surface defect detection using glam, Gabor wavelet, hog, and random forest classifier. Turkish J. Comput. Math. Educ. (TURCOMAT) **12**(12), 263–273 (2021)
14. Xu, H.J., Tang, Z.Q., Zhang, J.D., et al.: Research on mization of YOLOv5s detection algorithm for steel surface defect. Comput. Eng. Appl. **60**(07), 306–314 (2024)
15. Sohan, M., Sai Ram, T., Reddy, R., et al.: A review on yolov8 and its advancements. In: International Conference on Data Intelligence and Cognitive Informatics, pp. 529–545. Springer, Singapore (2024)
16. Zhang, X., Liu, C., Yang, D., et al.: RFAConv: Innovating Spatial Attention and Standard Convolutional Operation, pp 85–103. arxiv preprint arxiv:2304.03198 (2023)
17. Xia, Z., Pan, X., Song, S., et al.: Vision transformer with deformable attention. In: Proceedings of the IEEE/CVF Conference on Computer Vision and Pattern Recognition, pp 4794–4803 (2022)
18. Xiang, N., Gong, Z., Xu, Y., et al.: Material-aware path aggregation network and shape decoupled siou for x-ray contraband detection. Electronics **12**(5), 1179–1196 (2023)
19. Liu, C., Nong, Z.J., Zhang, Z.M.: Analysis of cold top forging cracking problem of cold heading steel wire rod and improvement measures. Southern Metal **06**(207), 30–33 (2015)

# Revised Local Laplacian Filter for Turbid Underwater Image Enhancement

**Shunsuke Takao and Taketsugu Hirabayashi**

**Abstract** Underwater optical images are of significance to measure detailed underwater environments and have attracted much attention. However, underwater images are degraded by physical processes such as absorption or scattering to blur image characteristics which impedes diverse image analysis. In particular, multiple scattering caused by turbid water reduces the visibility of underwater images to prevent efficient operations such as in marine construction or resource exploration. It is still challenging to enhance the turbid underwater images by means of physical models or deep learning; due to multiple scattering often simplified in physical models or less available data to train deep models. In this study, we propose a novel underwater image enhancement (UIE) method tailored for highly turbid underwater images. Our UIE method is built upon classic image processing techniques, Local Laplacian Filter (LLF) and Discrete Wavelet Transform (DWT), to effectively augment blurred image characteristics. Experiments on two UIE benchmark datasets demonstrate that the proposed *LLF-Fusion* produces the competitive performance in comparison to the other UIE techniques. We also conduct physical experiments with a water tank to verify the efficacy of the proposed method for 3D geometry reconstruction in turbid water.

**Keywords** Underwater image enhancement · Local Laplacian filter · Structure from motion · Marine construction

---

S. Takao (✉)
University of Tsukuba, Tsukuba, Japan
e-mail: takao.s.work@gmail.com

T. Hirabayashi
Port and Airport Research Institute, Yokosuka, Japan

 
Y. Ma (ed.), *Proceedings of 4th International Conference on Intelligent Systems Design and Engineering Applications*, Lecture Notes in Electrical Engineering 1478,
https://doi.org/10.1007/978-981-95-1802-9_13

## 1 Introduction

Underwater optical images are crucial to understand in-depth underwater environments and increasing attention has been attracted in marine research fields. They, however, subject to severe image degradation caused by complex underwater physical processes to impede detailed image analysis, such as 3D geometry estimation or object detection in underwater [1–3].

The signal attenuation significantly depends on wavelength in underwater to render underwater images with color cast in comparison to aerial images; for example, underwater images captured in open ocean showcase blueish appearances in contrast to greenish or yellowish colors observed in some turbid coastal regions [3–5]. Furthermore, scattering by suspended particles blur image characteristics in underwater medium; multiple scattering is more prominent in turbid water that is caused by gravel flung up by underwater construction machines in marine construction, reducing the contrast and visibility of underwater images (Fig. 1) [6, 7].

To effectively utilize underwater information, underwater image enhancement (UIE) is necessary to recover the above mentioned degradations, especially highly blurred details contained in turbid underwater images. UIE techniques have been developed from the viewpoints of signal processing, physical model, and deep learning. Classic signal processing techniques, such as Laplacian Pyramid [10] or Wavelet Transform [11], effectively augment high frequency signals by means of frequency decomposition to facilitate UIE [12–14]. However, they may amplify artifacts in a recovered image due to the spatially invariant linear processes [15]. Meanwhile, techniques based on underwater physical models [4] are not suitable to enhance the turbid underwater images due to multiple scattering in turbid water which is usually neglected to simplify the model [1, 3]. Even deep methods are less capable of enhancing the blurred details of hazy images, as is costly and sometimes impossible to collect many clear underwater images in turbid environments, which prohibits effective training of deep models [1, 16–18].

In this study, we propose a fusion-based UIE method, dubbed as *LLF-Fusion*, and verify the efficacy of the LLF-Fusion by conducting physical experiments with

**Fig. 1** Degraded underwater images (1st row) and recovered images processed by our method (2nd row). Our method, dubbed as *LLF-Fusion*, boosts feature extraction by Canny [8] (3rd column) and feature matching by SIFT [9] (4th column), while is failed in the degraded images

a water tank; our LLF-Fusion is tailored for underwater images of highly turbidity to enable applications such as leveling in marine construction or marine resource exploration. The proposed LLF-Fusion integrates frequency features that is built upon Local Laplacian Filter (LLF) [15] and Discrete Wavelet Transform (DWT) using Haar wavelet to enhance multiscale characteristics of hazy underwater images. In the proposed LLF-Fusion, an image is first input to be processed by LLF in three parameters to fuse sharpened and smoothed outputs of LLF. The sharpened result is transferred to the high frequency DWT components for image sharpening, while the smoothed LLF result, which is intuitively regarded as the backscattering component of an underwater image, is subtracted from the sharpened LLF result for color correction. Overview of our LLF-Fusion is shown in Fig. 2. Unlike classic Laplacian Pyramid that is implemented by spatially invariant Gaussian filter [10, 12], proposed LLF-Fusion incorporates LLF which enables edge-preserving image manipulations [15]. Thereby, we can augment high frequency signals effectively while avoiding halos for UIE. To the best of our knowledge, this is the first attempt to extend LLF to turbid underwater images.

Along with developing the LLF-Fusion, we conduct two kinds of physical experiments to examine the applicability of LLF-Fusion to marine construction. First, we take hazy pictures of a color chart by pouring milk to a small tank, to address the difficulty of collecting clear, ground-truth underwater images in real environments. We also measure the turbidity of the tank towards objective analysis. Second, we place a rubble mounds model in a water tank to reconstruct its 3D geometry via Structure from Motion (SfM) in turbid conditions by taking pictures from the surrounding viewpoints (Fig. 7); the 3D geometry estimation is necessary for leveling in marine construction [6].

Numerical experiments utilizing two benchmark datasets, which are taken in a small tank and real water areas, demonstrate that the proposed LLF-Fusion produces

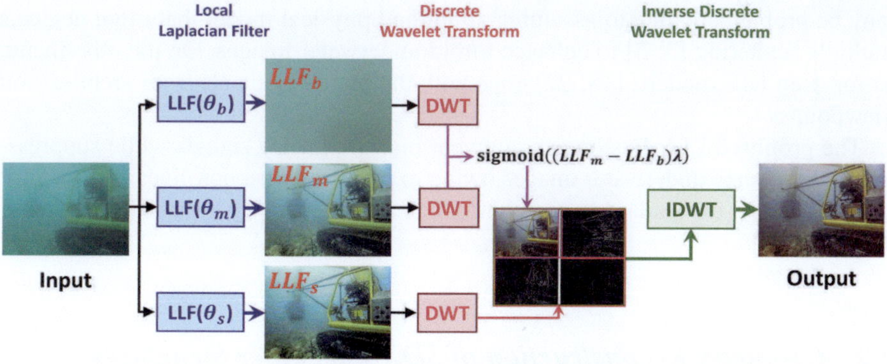


**Fig. 2** Overview of our LLF-Fusion. An input image is first processed by Local Laplacian Filter (LLF) [15] to produce three images, where smoothed and sharpened by different parameters of LLF. The LLF outputs are decomposed with Discrete Wavelet Transform (DWT) to fuse multiscale features for UIE

competitive performance in comparison to other UIE methods. We also verify that the proposed LLF-Fusion enables the 3D structure estimation of the rubble mounds model by SfM in turbid underwater, which leads to high resolution, cost-effective sensing in marine construction or marine resource exploration. Main contributions are summarized as follows.

- We propose a fusion based UIE method, *LLF-Fusion*, to enhance hazy underwater images by means of revised Local Laplacian Filter.
- We conduct physical experiments leveraging water tanks to evaluate our LLF-Fusion.
- Our LLF-Fusion produces competitive enhancement ability in comparison to other UIE techniques. We also verify that the LLF-Fusion contributes to underwater 3D reconstruction via SfM in turbid water conditions.

## 2 Related Work

### *2.1 Underwater Image Enhancement Methods*

To better cope with UIE, which is an inverse problem and challenging, some prior knowledge has been incorporated in the UIE literature to effectively enhance the severe image degradation. References [12–14] integrate classic signal processing techniques to decompose an image into low/high frequency components; blurred high frequency image characteristics are augmented by Laplacian Pyramid [12] or DWT [13, 14], whereas low frequency signals are processed by (revised) white balance methods to correct color cast. Other UIE methods have assumed underwater physical models [4, 19] or statistical properties, such as Dark Channel Prior [20, 21], to address biased color channels prominent in underwater images. Those approaches may be problematic due to assuming simplified physical model that often neglects multiple scattering [3, 5] to enhance turbid underwater images. On the other hand, as for deep UIE models, [18, 22] cope with the biased channels from architectural viewpoints.

The proposed LLF-Fusion can augment high frequency signals while suppressing halos of hazy underwater images owing to the pixel-wise non-linearities of LLF [15] in contrast to standard approaches which utilize spatially invariant linear operators [12–14].

### *2.2 Geometry Reconstruction of 3D Underwater Structures*

The 3D structure of underwater scenes has been successfully reconstructed by optical approaches such as Structure from Motion (SfM) [6, 23, 24], NeRF [2], and monocular depth estimation [25] to produce high resolution results; they are beneficially cost-effective in comparison to the other costly and cumbersome acoustic equipment or 3D laser scanners [24]. In particular, SfM can reconstruct the 3D structure of objects by mutually associating image features which are captured by different viewpoints. The issue of underwater SfM is caused by the degraded underwater image features which interfere with feature matching that is necessary for SfM [6, 23]; this problem is more serious in turbid water medium showing multiple scattering. In contrast to previous studies usually conducted in clear water conditions, we conduct SfM from turbid underwater images to explore the possibility to be applied to leveling in marine construction or marine resource exploration.

## 3 Method

We briefly introduce Local Laplacian Filter [15] (LLF) and Discrete Wavelet Transform (DWT) prior to the description of the proposed LLF-Fusion.

### *3.1 Local Laplacian Filter (LLF)*

Local Laplacian Filter (LLF) is developed by modifying Laplacian Pyramid [10] which enables edge-aware image manipulations by means of pixel-wise non-linearities [15]. Fast approximation algorithms are proposed in [26, 27].

Laplacian Pyramid is a classic signal processing method which hierarchically decomposes an image into pyramid representation [10, 15]. Each component of Laplacian Pyramid is usually computed by successively subtracting adjacent components of Gaussian Pyramid, which is obtained by spatially invariant Gaussian filter. The Laplacian Pyramid components are decomposed to features that reflect resolution or frequency of an image [10].

The steps of LLF is summarized as follows. Let $I_{in}$ be an input image. Here, Laplacian Pyramid and Gaussian Pyramid of $I_{in}$ is denoted as $L_l\left[I_{in}\right](p)$ and $G_l\left[I_{in}\right](p)$, respectively, where pyramid level $l$ and pixel $p$. The pyramid level $L$ and Gaussian kernel $G$ are given beforehand.

1. Build Gaussian Pyramid of $I_{in}$: $G_l\left[I_{in}\right](p) := v$
2. Apply remapping function $R_v$ to $I_{in}$, obtaining $I' = R_v(I_{in})$
3. Compute Laplacian Pyramid of $I'$: $L_l\left[I'\right](p) := w$
4. Substitute $w$ for output pyramid $L_l\left[I_{out}\right](p)$, for $p$ and $l$: $L_l\left[I_{out}\right](p) \leftarrow w$
5. Repeat phase 2 to 4 for all pixels $p$ and pyramid levels.
6. Compute output image $I_{out}$ by collapsing Laplacian Pyramid of $I_{out}$, obtaining $I_{out}$.

The above remapping function $R_v$ transforms an image depending on each pixel value $v$, which is defined for a target task such as detail manipulation or tone manipulation. To enhance details of blurred underwater images, the remapping function $R_{p_0}$ for pixel $p_0$ is defined as following equation [15].

$$R_{p_0}(i) = p_0 + sign(i - p_0) f(i)$$

$$f(i) = \begin{cases} \sigma_r \left(|i - p_0|/\sigma_r\right)^{\alpha} & if \quad |i - p_0| \leq \sigma_r \\ \sigma_r + \beta \left(|i - p_0| - \sigma_r\right) & otherwise \end{cases}$$

Here, $\alpha$, $\beta$, and $\sigma_r$ are user-defined hyper parameters. $\alpha$ is the parameter to control the effect of LLF, taking positive values. Detail enhancement is performed with $\alpha$ less than 1, while detail smoothing otherwise. $\sigma_r$ controls the strength of the manipulation, which distinguishes larger edges from finer details [15].

## 3.2 Discrete Wavelet Transform (DWT)

DWT is leveraged in many image manipulation tasks owing to the simple implementation with Haar Wavelet [11, 28]. Haar wavelet for image decomposition consists of following four kernels, $\{LL^T, LH^T, HL^T, HH^T\}$. Here, $L$ and $H$ are defined as $L^T := \frac{1}{\sqrt{2}}[1 \quad 1]$, $H^T := \frac{1}{\sqrt{2}}[-1 \quad 1]$, respectively. Note that $H$ gains high frequency components, while $L$ captures low frequency ones in a structurally organized manner. A divided image by DWT is reconstructed via Inverse Discrete Wavelet Transform (IDWT), which is the mirror operation of DWT, to reconstruct to the original resolution of an input image.

## 3.3 Scheme of Proposed LLF-Fusion

LLF enhances blurred structures of an image while preserving edges, depending on its hyper parameters, $\alpha$, $\beta$, $\sigma_r$, however, hyper parameter tuning is practically troublesome. Besides, color correction is necessary for UIE, therefore, we revise LLF to enhance underwater images effectively.

We first set fixed hyper parameters of LLF to enhance multiscale image features for UIE; $\theta_s$, $\theta_b$, and $\theta_m$ are set to produce different LLF outputs which are denoted as sharpened $LLF_s$, blurred $LLF_b$, and middle $LLF_m$. We then decompose $LLF_s$ via DWT inspired by [13, 14]. We directly adopt high frequency DWT components, $LLF_s$, $\{LH_s, HL_s, HH_s\}$, for a recovered image $I_{out}$ to enhance sharpness of an input image $I_{in}$. As for the LL component $LL_{out}$ of $I_{out}$, we apply following equation to remove color cast of underwater images.

$$LL_{out} = LL\left[sigmoid\left((LLF_m - LLF_b)\lambda\right)\right] \tag{1}$$

where $sigmoid$ represents sigmoid function and $\lambda$ is a scaling parameter. Here, $LL[\cdot]$ represents the low frequency $LL$ component decomposed by DWT. Finally,

**Fig. 3** Input images (2nd column), output images processed with our scheme (1st column). Results of LLF, $LLF_b$ (3rd column), $LLF_m$ (4th column), $LLF_s$ (5th column) obtained by setting LLF parameters $\theta = (\alpha, \sigma_r, \beta)$ as follows. $\theta_b = (3, 0.9, 1)$, $\theta_m = (0.25, 0.4, 1)$, and $\theta_s = (0.25, 0.6, 1)$

$I_{out}$ is reconstructed to the original resolution through IDWT based on $\{LL_{out}, LH_s, HL_s, HH_s\}$, which is expressed as:

$$I_{out} = IDWT\,[\{LL_{out}, LH_s, HL_s, HH_s\}] \tag{2}$$

Thus, multiscale image features are enhanced without tuning hyper parameters, promoting practical applications related to underwater images.

Figure 3 shows visual samples processed by the proposed LLF-Fusion and LLF. LLF-Fusion transforms input images (2nd column) to enhanced images (1st column) by fusing outputs of LLF, $LLF_b$, $LLF_m$, and $LLF_s$ (3rd to 5th column). In this processing, sharpness of land images is also enhanced with LLF-Fusion (2nd row), while LLF-Fusion is implemented for underwater images.

## 4 Numerical Experiments

### 4.1 Settings of Numerical Experiments

We evaluate the efficacy of the proposed LLF-Fusion for UIE tasks. We leverage Challenging-60 dataset [17] and our Turbid-Charts9, which contains hazy color chart images detailed in Sect. 5.1, to evaluate the performance of our LLF-Fusion. The Challenging-60 dataset contains 60 real underwater images of severe degradation and hence selected [17]. We adopt Peak Signal-to-Noise Ratio (PSNR) and Structural Similarity Index Measure (SSIM) for Turbid-Charts9, while UIQM [29] and UCIQE [30] are compared for Challenging-60, as no ground truths are usually available in real environments. We empirically set hyper parameters of LLF-Fusion $\theta = (\alpha, \sigma_r, \beta)$ as follows and fixed throughout experiments; $\theta_s = (0.25, 0.6, 1)$, $\theta_m = (0.25, 0.4, 1)$, $\theta_b = (3, 0.9, 1)$, $\lambda = 5$.

## *4.2 Comparison to Other UIE Methods*

**Results of Challenging-60** We compare the proposed LLF-Fusion with the following UIE methods in this experiment; Fusion [12], PUIE-Net [16], Water-Net [17], U-Trans [18], and DICAM [22].

The performance comparisons are illustrated in Table 1 and Fig. 4. Our LLF-Fusion produces the best UIQM [29] and UCIQE [30] in the experiment. From visual comparison in Fig. 4, the proposed method enhances degraded images and vivid results are obtained (2nd row); various color appearances are corrected by LLF-Fusion and sharpness and contrast are effectively improved. Here, we balance the intensity of R channel as a preprocessing [12] to adapt to wavelength-selective signal attenuation in underwater. In unsupervised Fusion [12] (3rd row), saturation is lost and blurred due to the severe degradation of the original images. Water-Net [17] (5th row), which fuses outputs processed by classic signal processing methods to normalize inputs, largely correct color distortion, however, sharpness is insufficiently improved. U-Trans [18] (6th row) and PUIE-Net [16] (4th row) produce a little hazy outputs. DICAM [22] cannot restore yellowish color cast (7th row).

**Results of Turbid-Charts9** We adopt Fusion [12], U-Trans [18], DW-GAN [28], Histogram Equalization (HE), and Dark Channel Prior (DCP) [20] for baseline methods; we focus on underwater images of severe turbidity and aim to compare methods from broader fields in this experiment.

Qualitative and quantitative results are shown in Fig. 5, Tables 2, and 3. As illustrated in Tables 2 and 3, our method performs better than previous methods in Turbid-Charts9; our LLF-Fusion achieves best 8 out of 9 in PSNR, whilst 6 out of 9 in SSIM. For visual comparison in Fig. 5, LLF-Fusion improves the visibility of foggy raw images, clarifying the contours until the case of 7 mL (6th column). On the other hand, U-Trans [18] fails to reconstruct overall color tones due to the severe turbidity (7th row). Fusion [12] amplifies fine artifacts caused by its linear processes. DCP [20] performs well since the input color cast is less, but luminance is concurrently lost. Classic HE insufficiently improves contrast of images. DW-GAN [28], terrestrial dehazing method, produces serious color bias originated from domain shift. Overall, all methods are less capable of enhancing the severe turbidity in Turbid-Charts9, due to the difficulty of inverse problems and is marked as the future work (7th to 9th columns of Fig. 5).

**Table 1** UIQM and UCIQE in Challenging-60 dataset

| Metric/Method | Propose | Fusion | PUIE-Net | Water-Net | U-Trans | DICAM |
|---|---|---|---|---|---|---|
| UIQM | 2.909 | 2.770 | 2.752 | 2.609 | 2.724 | 2.722 |
| UCIQE | 5.131 | 3.768 | 3.790 | 4.554 | 4.231 | 4.486 |

Best score is marked in red

**Fig. 4** Real underwater images from Challenging-60 [17] (1st row). Recovered results of our LLF-Fusion (2nd row), Fusion [12] (3rd row), PUIE-Net [16] (4th row), Water-Net [17] (5th row), U-Trans [18] (6th row), DICAM [22] (7th row) are shown

## 4.3 Insights on LLF-Fusion and Ablated Results

We explain roles of LLF and Eq. 1 in our LLF-Fusion.

In LLF, we observe that edges of a turbid image are more captured in $HH_s$, $HL_s$, $LH_s$ decomposed by DWT, whereas do not appear in $HH$, $HL$, $LH$ produced from a raw image. Meanwhile, no salient difference is illustrated in the clear image. From these observation, we can augment high frequency signals of turbid images by just substituting $LH_s$, $HL_s$, $HH_s$ instead of the original components.

**Fig. 5** Visual comparison for our Turbid-Charts9 taken with increased turbidity. Recovered results of our method (2nd row), Fusion [12] (3rd row), DCP [20] (4th row), HE (5th row), DW-GAN [28] (6th row), and U-Trans [18] (7th row) are shown

**Table 2** PSNR in Turbid-Charts9

| Milk (ml)/Turbidity | 0 | 3/26 | 4/28 | 5/32 | 6/35 | 7/38 | 8/44 | 9/46 | 10/51 |
|---|---|---|---|---|---|---|---|---|---|
| Propose | 20.502 | 18.372 | 15.885 | 15.840 | 14.730 | 14.137 | 14.437 | 13.777 | 13.576 |
| Fusion | 25.385 | 17.735 | 13.625 | 14.120 | 13.035 | 12.407 | 13.353 | 12.344 | 11.892 |
| DCP | 12.298 | 12.716 | 13.589 | 14.037 | 13.350 | 12.928 | 14.197 | 13.526 | 13.285 |
| HE | 21.322 | 14.462 | 13.406 | 13.078 | 12.264 | 12.417 | 11.641 | 12.423 | 11.659 |
| DW-GAN | 15.555 | 15.441 | 12.336 | 11.238 | 10.669 | 10.438 | 9.802 | 10.341 | 9.703 |
| U-Trans | 25.483 | 14.062 | 13.476 | 13.089 | 12.920 | 12.479 | 13.103 | 11.629 | 11.706 |

Best score is marked in red

We formulate Eq. 1 to correct color cast of underwater images, which are often modeled as summation of a signal term and a veiling light term [3]. We empirically found that $LLF_b$ approximately corresponds to the veiling light of the degraded underwater images and we can correct color cast by subtracting $LLF_b$ from $LLF_m$; that is inspired from [31] pointing out that the top Laplacian Pyramid component is similar to the veiling light. The sigmoid function in Eq. 1 suppresses saturation of RGB values of an image. In Fig. 6, the severe color cast is largely removed in the proposed scheme.

**Table 3** SSIM in Turbid-Charts9

| Milk (ml)/Turbidity | 0 | 3/26 | 4/28 | 5/32 | 6/35 | 7/38 | 8/44 | 9/46 | 10/51 |
|---|---|---|---|---|---|---|---|---|---|
| Propose | 0.725 | 0.696 | 0.654 | 0.646 | 0.591 | 0.600 | 0.541 | 0.593 | 0.532 |
| Fusion | 0.935 | 0.602 | 0.445 | 0.354 | 0.284 | 0.261 | 0.219 | 0.187 | 0.140 |
| DCP | 0.666 | 0.549 | 0.506 | 0.494 | 0.439 | 0.440 | 0.384 | 0.389 | 0.346 |
| HE | 0.841 | 0.659 | 0.616 | 0.596 | 0.574 | 0.575 | 0.554 | 0.573 | 0.553 |
| DW-GAN | 0.752 | 0.558 | 0.488 | 0.465 | 0.407 | 0.432 | 0.358 | 0.401 | 0.354 |
| U-Trans | 0.911 | 0.530 | 0.461 | 0.380 | 0.359 | 0.333 | 0.401 | 0.224 | 0.350 |

Best score is marked in red

**Fig. 6** Results of $LLF_m - LLF_b$ of Eq. 1 (left). Luminance is manually adjusted for visualization (right)

## 5 Physical Experiments

We conduct physical experiments utilizing water tanks to collect hazy underwater images (Sect. 5.1) and underwater SfM (Sect. 5.2) towards applications for marine construction or resource exploration.

### 5.1 Collection of Hazy Pictures of Color Chart

We take pictures of a color chart in turbid water medium, dubbed as *Turbid-Charts9*, to evaluate UIE methods with ground truths. The pictures are taken from distance about 30 cm long by GoPro HERO9; the visual images are presented in Fig. 5. In this physical experiment, we utilize a WDKK Waterproof Color Chart from DGK Color Tools and place it to a small water tank. We pour milk at intervals of 1 mL to increase the turbidity, adding 3–10 milliliters of milk. The visibility of the color chart is decreased from the original image (1st column) with increased amount of added milk (Fig. 5). e measure the turbidity with a turbidimeter, Digital turbidity 500 in total Model DPM2-TB500 from KYORITSU CHEMICAL-CHECK Lab.,Corp. As shown in 1st row of Table 2, turbidity ranges from 26 to 51 in this measurement.

## 5.2 Underwater Structure from Motion in Turbid Water

The 3D geometry of underwater rubble mounds is necessary to confirm whether the mounds is sufficiently flat for leveling in marine construction [6]. However, the severe turbidity, which is caused by gravel flung up by construction machines in underwater, disturbs optical sensing like SfM (Fig. 1) [6].

Motivated by this, we explore the possibility to reconstruct the 3D geometry of underwater scenes in turbid conditions. As shown in 1st column of Fig. 7, we first simulate the state of turbidity by filling up muddy water to a housing case for a camera. Subsequently, we place the rubble mounds model in a water tank and then take the pictures from its surrounding viewpoints, to reconstruct the 3D structure via SfM. More specifically, we conduct following procedures to collect hazy underwater images.

1. We dissolve kaoline as suspended solid and pour into a housing case which is mounted to a camera.
2. We sink a rubble mounds model having rocks to a water tank. We also place a color chart and measures.
3. We take underwater images of the rubble mounds model from the top view by GoPro HERO9 which is mounted to BlueROV2.

In conducting SfM, we utilize Agisoft Metashape version 1.8.2 (Agisoft LLC., St. Petersburg, Russia) to reconstruct the 3D structure of the rubble mounds model; we resize the underwater images to 996 × 1328 from 3984 × 5312 height and width to save computational time.

Figure 7 illustrates visual results of SfM. In Fig. 7, the captured underwater images are dark and hazy, failing to reconstruct the 3D geometry due to the degraded image characteristics. Meanwhile, we can enhance the degraded features by means of our LLF-Fusion to produce the textured model by SfM. Thereby, we can largely capture the configuration of the rocks via SfM in turbid water. The optical approaches as presented can measure the high resolution, colored 3D geometry in low cost, thus would be fitted for widespread applications in comparison to acoustic equipment [23, 24].

**Fig. 7** Experiments with a water tank for SfM in turbid underwater. We add Kaoline to a housing case of a camera and take hazy pictures by GoPro HERO9. The feature matching in SfM is failed in the degraded underwater images (2nd column), while is succeeded to reconstruct the 3D geometry enhanced by our LLF-Fusion (3rd column)

## 6 Conclusion

We proposed a fusion-based UIE method, dubbed as *LLF-Fusion*, to effectively enhance turbid underwater images. The proposed LLF-Fusion is developed with Local Laplacian Filter (LLF) to boost blurred image characteristics in an edge-preserving manner in contrast to previous linear operations which sometimes produce serious halos. Numerical experiments on two hazy underwater image datasets demonstrate the efficacy of our LLF-Fusion for UIE. We conduct physical experiments as well to reconstruct the 3D geometry of the rubble mounds model via SfM in turbid water, which facilitates cost-effective measurement in marine construction.

## References

1. Anwar, S., Li, C.: Diving deeper into underwater image enhancement: a survey. Signal Process. Image Commun. **89**, 115978 (2020). https://www.sciencedirect.com/science/article/pii/S0923596520301478
2. Levy, D., Peleg, A., Pearl, N., Rosenbaum, D., Akkaynak, D., Korman, S., Treibitz, T.: Seathru-nerf: neural radiance fields in scattering media. In: Proceedings of the IEEE/CVF Conference on Computer Vision and Pattern Recognition, pp. 56–65 (2023)
3. Akkaynak, D., Treibitz, T.: A revised underwater image formation model. In: Proceedings of the IEEE Conference on Computer Vision and Pattern Recognition (CVPR) (2018)
4. Akkaynak, D., Treibitz, T.: Sea-thru: a method for removing water from underwater images. In: Proceedings of the IEEE/CVF Conference on Computer Vision and Pattern Recognition (CVPR) (2019)
5. Takao, S.: Underwater image sharpening and color correction via dataset based on revised underwater image formation model. In: The Visual Computer, pp. 1–16 (2024)
6. Takao, S., Kita, T., Hirabayashi, T.: Towards reconstruction of 3d geometry of underwater rubble mounds via structure from motion. In: 2023 IEEE Underwater Technology (UT), pp. 1–6 (2023)
7. Duarte, A., Codevilla, F., Gaya, J.D.O., Botelho, S.S.: A dataset to evaluate underwater image restoration methods. In: OCEANS 2016-Shanghai, pp. 1–6. IEEE (2016)
8. Canny, J.: A computational approach to edge detection. IEEE Trans. Pattern Anal. Mach. Intell. **6**, 679–698 (1986)
9. Lowe, D.G.: Distinctive image features from scale-invariant keypoints. Int. J. Comput. Vis. **60**(2), 91–110 (2004)
10. Burt, P.J., Adelson, E.H.: The Laplacian pyramid as a compact image code. In: Readings in Computer Vision, pp. 671–679. Elsevier (1987)
11. Yoo, J., Uh, Y., Chun, S., Kang, B., Ha, J.W.: Photorealistic style transfer via wavelet transforms. In: Proceedings of the IEEE/CVF International Conference on Computer Vision, pp. 9036–9045 (2019)
12. Ancuti, C.O., Ancuti, C., De Vleeschouwer, C., Bekaert, P.: Color balance and fusion for underwater image enhancement. IEEE Trans. Image Process. **27**(1), 379–393 (2017)
13. Zhou, J., Wei, X., Shi, J., Chu, W., Lin, Y.: Underwater image enhancement via two-level wavelet decomposition maximum brightness color restoration and edge refinement histogram stretching. Opt. Express **30**(10), 17290–17306 (2022)
14. Rowghanian, V.: Underwater image restoration with haar wavelet transform and ensemble of triple correction algorithms using bootstrap aggregation and random forests. Sci. Rep. **12**(1), 8952 (2022)

15. Paris, S., Hasinoff, S.W., Kautz, J.: Local Laplacian filters: edge-aware image processing with a Laplacian pyramid. ACM Trans. Graph. **30**(4), 68 (2011)
16. Fu, Z., Wang, W., Huang, Y., Ding, X., Ma, K.K.: Uncertainty inspired underwater image enhancement. In: European Conference on Computer Vision, pp. 465–482. Springer (2022)
17. Li, C., Guo, C., Ren, W., Cong, R., Hou, J., Kwong, S., Tao, D.: An underwater image enhancement benchmark dataset and beyond. IEEE Trans. Image Process. **29**, 4376–4389 (2019)
18. Peng, L., Zhu, C., Bian, L.: U-shape transformer for underwater image enhancement. IEEE Trans. Image Process. **32**, 3066–3079 (2023)
19. Zhang, Z., Jiang, Z., Liu, J., Fan, X., Liu, R.: Waterflow: heuristic normalizing flow for underwater image enhancement and beyond. In: Proceedings of the 31st ACM International Conference on Multimedia, pp. 7314–7323 (2023)
20. He, K., Sun, J., Tang, X.: Single image haze removal using dark channel prior. IEEE Trans. Pattern Anal. Mach. Intell. **33**(12), 2341–2353 (2010)
21. Peng, Y.T., Cao, K., Cosman, P.C.: Generalization of the dark channel prior for single image restoration. IEEE Trans. Image Process. **27**(6), 2856–2868 (2018)
22. Tolie, H.F., Ren, J., Elyan, E.: Dicam: deep inception and channel-wise attention modules for underwater image enhancement. Neurocomputing **584**, 127585 (2024)
23. Young, G., Dey, S., Rogers, A., Exton, D.: Cost and time-effective method for multi-scale measures of rugosity, fractal dimension, and vector dispersion from coral reef 3d models. PloS One **12**(4), e0175341 (2017)
24. Massot-Campos, M., Oliver-Codina, G.: Optical sensors and methods for underwater 3d reconstruction. Sensors **15**(12), 31525–31557 (2015)
25. Zhang, F., You, S., Li, Y., Fu, Y.: Atlantis: enabling underwater depth estimation with stable diffusion. In: Proceedings of the IEEE/CVF Conference on Computer Vision and Pattern Recognition, pp. 11852–11861 (2024)
26. Aubry, M., Paris, S., Hasinoff, S.W., Kautz, J., Durand, F.: Fast local Laplacian filters: theory and applications. ACM Trans. Graph. (TOG) **33**(5), 1–14 (2014)
27. Sumiya, Y., Otsuka, T., Maeda, Y., Fukushima, N.: Gaussian Fourier pyramid for local Laplacian filter. IEEE Signal Process. Lett. **29**, 11–15 (2021)
28. Fu, M., Liu, H., Yu, Y., Chen, J., Wang, K.: Dw-gan: a discrete wavelet transform gan for nonhomogeneous dehazing. In: Proceedings of the IEEE/CVF Conference on Computer Vision and Pattern Recognition, pp. 203–212 (2021)
29. Panetta, K., Gao, C., Agaian, S.: Human-visual-system-inspired underwater image quality measures. IEEE J. Ocean. Eng. **41**(3), 541–551 (2015)
30. Yang, M., Sowmya, A.: An underwater color image quality evaluation metric. IEEE Trans. Image Process. **24**(12), 6062–6071 (2015)
31. Takao, S.: Zero-shot image enhancement with renovated Laplacian pyramid. In: Karlinsky, L., Michaeli, T., Nishino, K. (eds.) Computer Vision—ECCV 2022 Workshops, pp. 721–737. Springer Nature Switzerland, Cham (2023)

# Lightweight UAV-Based Road Disease Detection Algorithm

**Guo Jinbo, Wang Shenghuai, Xu Fenghua, Chen Xiaohui, Wang Chen, and Zhang Wei**

**Abstract** To enable efficient operation of drone detection models on resource-constrained devices while maintaining high accuracy, an improved lightweight UAV pavement damage detection algorithm, named LHP-YOLO, is proposed. This algorithm enhances YOLOv8n by focusing on two aspects: lightweight design and improved detection accuracy. First, the RG-C2f module is introduced to replace the original C2f module in YOLOv8, reducing redundant computations in the model. Additionally, a lightweight detection head, Detect-T3G, is designed to minimize the number of parameters. To further improve the model's performance in complex backgrounds and small object detection, Content-Guided Attention Fusion (CGA Fusion) is incorporated into the feature fusion structure of the YOLOv8n neck network. Moreover, Channel-Wise Knowledge Distillation (CWD) is applied to boost the accuracy of the lightweight model. Experimental results on the publicly available pavement damage dataset RDD 2022 show that, compared to the original YOLOv8n model, LHP-YOLO achieves a 3.4% increase in mAP50 while reducing the model size and computation by 41.3% and 51.8%, respectively, with parameters totaling 1.76 M and 3.9 GFLOPs. These improvements meet the real-time performance requirements for UAV-based pavement damage detection.

**Keywords** UAV · Pavement disease · Object detection · YOLOv8 · Lightweight · Knowledge distillation

## 1 Introduction

UAV-based pavement damage detection focuses on the localization and fine-grained classification of road defects from images. With the rapid expansion of rural road construction in China, road network density and traffic demands have significantly

G. Jinbo · W. Shenghuai (✉) · X. Fenghua · C. Xiaohui · W. Chen · Z. Wei
School of Automotive Intelligent Manufacturing, Hubei University of Automotive Technology, Shiyan, China
e-mail: shwangkb@163.com

 
Y. Ma (ed.), *Proceedings of 4th International Conference on Intelligent Systems Design and Engineering Applications*, Lecture Notes in Electrical Engineering 1478,
https://doi.org/10.1007/978-981-95-1802-9_14

increased, making the maintenance and management of rural roads critical for traffic flow and regional economic development. Traditionally, rural road maintenance has relied on manual inspections, which are inefficient and subjective, leading to inconsistent and inaccurate assessments. Therefore, the ability to quickly and accurately detect pavement damage in complex road environments is vital for traffic safety and public security. Compared to traditional detection methods, UAV-based detection [1] has several unique characteristics, including the ability to capture targets from high altitudes at various angles while distinguishing objects in complex environments. However, the algorithm must be highly real-time, lightweight, and power-efficient to function effectively on embedded devices with limited resources, ensuring extended flight times and efficient task execution.

Object detection, as one of the core technologies in UAV detection, typically involves two main categories of algorithms: two-stage detection algorithms based on candidate regions and one-stage detection algorithms based on regression. Two-stage object detection algorithms, such as Faster R-CNN [2], Cascade R-CNN [3], and R-FCN [4], excel in high accuracy and performance in complex scenes. In contrast, one-stage object detection algorithms, including YOLO [5], SSD [6], EfficientDet [7], and DETR [8], are renowned for their speed and real-time performance. Given the specific characteristics of UAV object detection, researchers predominantly employ one-stage detection algorithms for this task due to their efficiency and applicability in real-time scenarios. In [9], a lightweight object detection algorithm based on an improved YOLOv5s is proposed, which integrates Ghost modules, coordinate attention mechanisms, and simplified spatial pyramid pooling to achieve a reduction in model parameters and computation while improving detection accuracy. The work in [10] introduces convolutional block attention modules, deep convolution, and knowledge distillation techniques, significantly enhancing small object detection performance. In [11], the introduction of C3Ghost and GhostConv modules into the YOLOv5 backbone, combined with the DWConv module and C3Ghost in the neck network, and the addition of coordinate attention mechanisms, effectively reduces the model size and enhances detection speed. Despite these advancements, one-stage detection algorithms, which perform classification and regression directly on the feature map without generating candidate regions, struggle with small targets and complex backgrounds. This often results in missed detections and false positives, underscoring the need for further improvements in detecting small objects in UAV-based systems.

In the research on improving model accuracy, distillation techniques have demonstrated a significant role in enhancing the performance of object detection models. In [12], a novel Localization Distillation method is proposed, which focuses on estimating the localization quality and addressing the issue of rotating object detection. This approach significantly improves the localization accuracy of object detection. Additionally [13] introduces a Label Assignment Distillation method, where collaborative learning of label assignment enhances the training effectiveness of the object detection network. In summary, current research on balancing accuracy and lightweight design primarily focuses on network architecture and distillation techniques, but few studies integrate both. Bridging this gap could further enhance model performance and efficiency.

This paper proposes LHP-YOLO, an efficient UAV-based pavement damage detection algorithm, built on YOLOv8n with a focus on lightweight design and distillation techniques. Key improvements include replacing the C2f module with RG-C2f to reduce redundant computations while maintaining accuracy, introducing the Detect-T3G lightweight detection head to lower model parameters, incorporating Content-Guided Attention Fusion (CGA Fusion) in the neck network to enhance small object detection by dynamically adjusting feature weights, and applying Channel-Wise Knowledge Distillation (CWD) to further refine accuracy without increasing complexity. These enhancements make LHP-YOLO both efficient and effective for UAV-based applications.

## 2 YOLOv8 Algorithm

YOLOv8 is a versatile algorithm integrating object detection, image segmentation, and classification. Its architecture consists of four key components: the input layer, backbone network, neck network, and detection head. Featuring a redesigned backbone and neck (including the C2f module) and an anchor-free mechanism, YOLOv8 improves detection accuracy while simplifying training. It supports dynamic input sizes and multi-scale detection, allowing effective handling of varying object sizes. The optimized detection head enhances both speed and accuracy, while built-in training tools like mixed-precision training and advanced data augmentation improve adaptability across hardware environments. Balancing speed, accuracy, and flexibility, YOLOv8 is well-suited for real-time tasks, including UAV-based detection applications.

YOLOv8 has achieved significant advantages in terms of speed, accuracy, and usability. It offers various model sizes, such as YOLOv8n, YOLOv8s, and others, allowing users to balance performance and computational resources based on their specific needs. Thanks to its exceptional real-time performance and flexibility, YOLOv8 is widely applied in diverse fields, including autonomous driving, medical image analysis, retail data monitoring, and road defect detection. As a result, YOLOv8 has become a powerful tool for both industry research and practical applications. Its ability to handle a wide range of tasks efficiently, while maintaining high levels of accuracy, makes it a go-to solution for various real-time detection challenges across multiple domains.

# 3 LHP-YOLO Algorithm

## 3.1 Lightweight Network Architecture Design

Rural road pavement disease detection requires identifying various surface defects while ensuring both accuracy and efficiency. To address challenges such as low localization accuracy for small targets, poor real-time performance, and resource constraints in UAV-based detection, a lightweight YOLOv8n-based model, LHP-YOLO, is proposed. Key improvements include replacing the C2f module with RG-C2f to reduce redundant calculations while maintaining detection performance, designing the Detect-T3G head to lower parameter count and memory usage, and incorporating Content-Guided Attention Fusion (CGA Fusion) into the neck network to enhance small target detection in complex backgrounds. These optimizations make LHP-YOLO both efficient and accurate, meeting real-time UAV-based pavement defect detection requirements in rural environments. The LHP-YOLO network structure is shown in Fig. 1.

**CSPRGX Network Architecture**. Due to the frequent changes in flight altitude and angles of UAVs, the captured images often exhibit varying perspectives and scales. The C2f module in YOLOv8, based on the ELAN structure from YOLOv7, struggles to capture target variations across all scales in rapidly changing UAV-captured scenes and suffers from large parameter counts. To address these issues, this paper introduces the RG-C2f module, constructed using a residual structure composed of Re-parameterized Refocusing Convolution (RepConv) [14], standard convolution, and auxiliary convolution, as shown in Fig. 2. This RG-C2f module replaces the original C2f in the network, resulting in a new network structure named CSPRGX. Drawing inspiration from GhostNet, RG-C2f uses inexpensive one-dimensional auxiliary

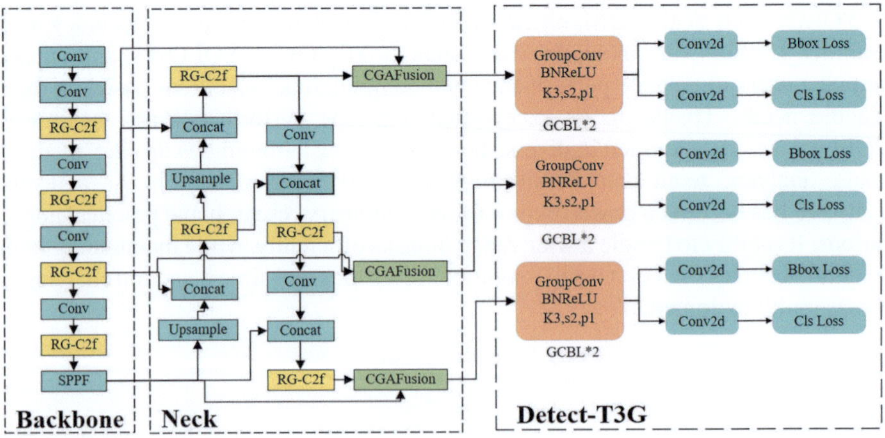


**Fig. 1** LHP-YOLO network structure diagram

convolutions to generate redundant feature maps, thereby addressing the widespread redundancy found in intermediate feature maps from mainstream CNNs. This reduces both computational load and parameter count. The RepConv structure employs a multi-branch architecture during training, consisting of one standard convolution layer and several auxiliary convolution layers. During inference, the multi-branch convolution layers are merged into a single convolution layer. Compared to standard convolution, RepConv achieves higher accuracy with lower computational cost and memory usage. The CSPRGX network effectively extracts and fuses features, allowing it to detect pavement defects captured from different angles and scales, while also reducing redundant computations and access costs, ultimately achieving network lightweighting.

**Detect-T3G Detection Head**. Real-time performance is critical in UAV-based pavement defect detection to enable road maintenance departments to make timely decisions and arrange repairs quickly. However, UAVs often face limitations in computational power and storage, restricting the use of complex algorithms. YOLOv8's detection head features a decoupled structure, separating object classification and bounding box regression into two independent branches. While this improves detection accuracy, the use of separate components for each task increases the network's complexity, raising computational overhead and affecting real-time performance. This added complexity can reduce the system's effectiveness in scenarios requiring fast decision-making, especially on resource-constrained UAVs.

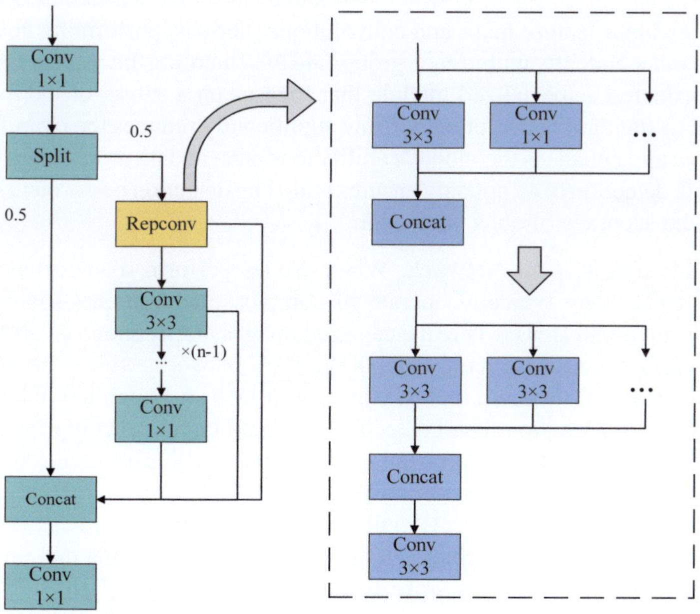


**Fig. 2** RG-C2f structure diagram

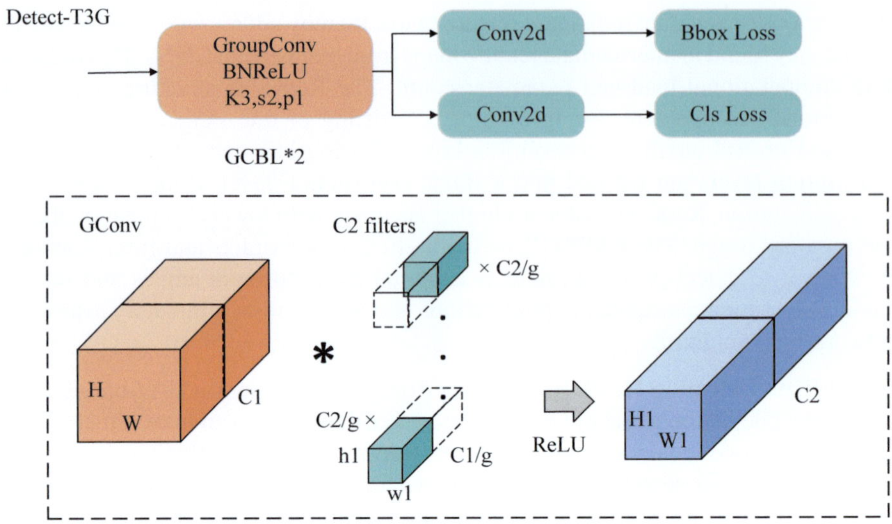


**Fig. 3** Detect-T3G structure diagram

To address the aforementioned issues, this paper optimizes the detection head of YOLOv8 by simplifying and merging the standard convolutions in the decoupled head structure, replacing them with Grouped Convolution (GConv) [15], as shown in Fig. 3. The new detection head structure is named Detect-T3G. GConv operates by grouping the input feature maps and convolutional kernels, performing convolution operations independently within each group, and then merging the results. Each group can be considered a specialized module that focuses on a subset of the input data. Compared to standard convolution, GConv significantly reduces the overall computational cost and enhances the model's ability to represent data more effectively. The Detect-T3G detection head not only ensures real-time detection performance but also improves the accuracy of object detection.

**Efficient Feature Fusion Network**. When drones perform road surface damage collection tasks, they typically operate in complex environments such as cities, forests, and suburban areas, where the captured images often contain small and dense targets. YOLOv8's neck structure adopts the PAN network, which can effectively fuse feature maps of different scales. However, it lacks a design to enhance image details, which may lead to missed detections of small targets in complex scenes. To improve the model's ability to detect detailed information, this paper introduces CGA Fusion [16] into the feature fusion network of YOLOv8's neck structure, as shown in Fig. 4. CGA Fusion first extracts multi-level features from the image, then uses a content-guided mechanism to obtain global information. Following this, an attention mechanism dynamically weighs the features, and finally, these weighted features are fused across multiple levels, integrating low-level details with high-level semantic information. Compared to the original YOLOv8 neck structure, the efficient feature fusion network with CGA Fusion can emphasize important regions and suppress

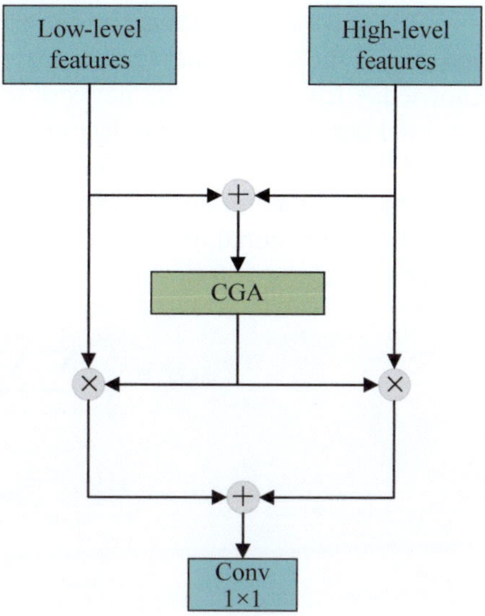


**Fig. 4** CGA fusion structure diagram

irrelevant parts, resulting in detailed and clear images that enhance the accuracy of object detection.

## 3.2 *Knowledge Distillation*

Through lightweight processing, the model's computational complexity and parameter count have been reduced, but detection accuracy saw limited improvement. To ensure the UAV-based road surface damage detection system meets deployment requirements for resource-constrained devices while maintaining high accuracy, this paper applies Channel-Wise Knowledge Distillation (CWD). CWD transfers knowledge from a teacher model to a student model at the channel level, helping the student model focus on the most informative features. By emphasizing specific channel-wise feature representations, CWD enables the lightweight model to maintain high performance without a significant increase in computational cost. This approach enhances the model's ability to detect road surface damage accurately, even on devices with limited computational resources.

**Channel Level Knowledge Distillation**. Knowledge distillation is a technique that transfers the knowledge learned by a large and complex neural network (referred to as the teacher network) to a smaller, simpler network (referred to as the student network). This allows the student network to achieve improved performance while retaining its reduced computational resource requirements. In this paper, CWD [17] is employed to enhance the accuracy of the lightweight model for drone-based road surface damage detection.

CWD specifically focuses on the channel-level features of the teacher network and guides the student network to learn from these key features. By leveraging the knowledge from the teacher network in a more compact form, the student network can maintain high detection accuracy, even in resource-constrained environments typical of drone applications. This approach significantly improves the performance of the model, making it more effective for real-time, on-site damage detection with limited computational resources.

The formula is as follows:

$$L_{\text{distill}} = \frac{T^2}{C}\sum_{c=1}^{C}\sum_{i=1}^{WH}\varnothing\left(y^T_{C,i}\right)\log\left[\frac{\varnothing\left(y^t_{c,i}\right)}{\varnothing\left(y^s_{c,i}\right)}\right] \quad (1)$$

$$\varnothing(y_C) = \frac{\exp\left(\frac{y_{c,i}}{T}\right)}{\sum_{i=1}^{WH}\exp\left(\frac{y_{c,i}}{T}\right)} \quad (2)$$

$$L_{\text{total}} = L_{\text{train}} + \alpha L_{\text{distill}} \quad (3)$$

$t$ is the teacher network; $s$ is a student network; $C$ is the channel index; $i$ is the spatial position of the corresponding channel; $y^t_{c,i}$ and $y^s_{c,i}$ represent the eigenvalues of the teacher model and student model at $C$ channels and $i$ positions, respectively; $\varnothing$ is a normalization function used to normalize the eigenvalues; $T$ is the distillation temperature, with a default value of 1; $L_{\text{train}}$ is the training model loss; $L_{\text{distill}}$ is the loss of knowledge distillation; $\alpha$ is the proportion of distillation loss.

### *3.3 Dataset and Experimental Setup*

The China Drone dataset [18] from the RDD 2022 is used in this study to validate the effectiveness of the proposed model. The RDD 2022 dataset contains 47,420 road images, capturing over 55,000 road damage events from six different countries: Japan, India, Czech Republic, Norway, the United States, and China. These images were taken in diverse environments and under various conditions, covering six major types of road damage: longitudinal cracks (D00), transverse cracks (D10), mesh cracks (D20), patches (Repair), block cracks (Block crack), and potholes (D40). Examples of typical road surface damage captured by drones are shown in Fig. 5. For the purposes of this research, the dataset was divided into three subsets: a training set, a validation set, and a test set, with a distribution ratio of 7:1:2.

To reduce overfitting, boost model robustness, and enhance performance on unseen data, this study applies several online data augmentation techniques. These include using filters with varying kernel sizes to blur images, converting RGB images to grayscale and inverting them when the average pixel value exceeds 127, and applying Contrast-Limited Adaptive Histogram Equalization (CLAHE) to enhance local contrast. Random adjustments to brightness and contrast, along with gamma

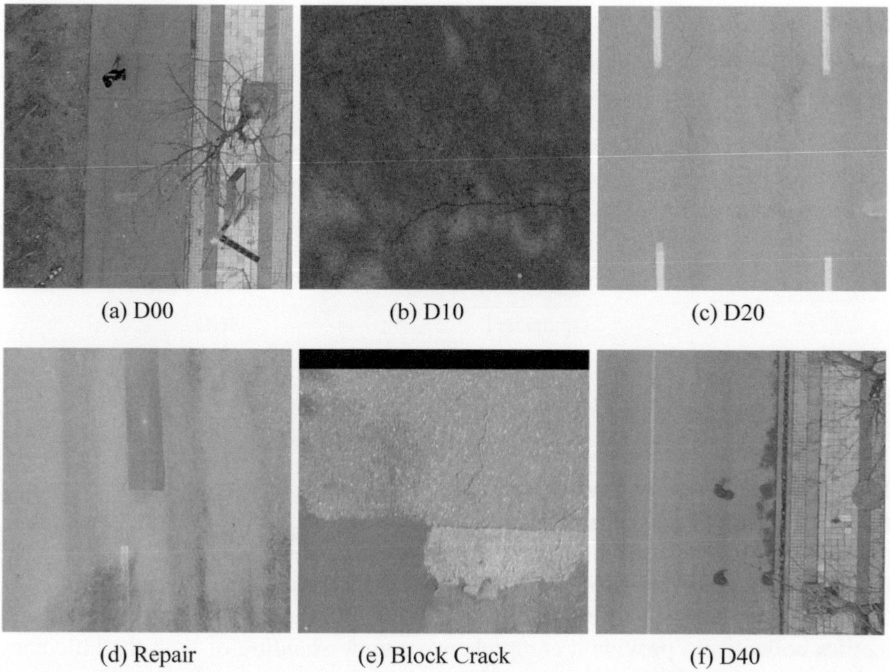


**Fig. 5** 6 typical road surface diseases captured by drones

correction to simulate exposure changes, are incorporated. Additionally, Joint Photographic Experts Group (JPEG) and Web Picture (WebP) compression techniques simulate data compression effects, enriching the training dataset. These strategies improve the model's adaptability and generalization, enhancing its ability to perform effectively on new data.

The hardware platform and environmental parameters used during the experimental training phase are shown in Table 1.

The main parameters adjusted during the model training phase are shown in Table 2.

**Table 1** Training environment and hardware platform parameter table

| Parameter | Font size and style |
|---|---|
| CPU | Intel (R) Xeon (R) Gold 6430 |
| GPU | RTX 4090 |
| GPU memory size | 24 GB |
| Operating system | Win 11 |
| Deep learning architecture | PyTorch2.0.0 + Python3.8 + Cuda 11.8 |

**Table 2** Experimental parameter settings

| Parameter | Figure |
|---|---|
| Number of iterations | 300 |
| Momentum factor | 0.937 |
| Initial learning rate | 0.01 |
| Final learning rate | 0.01 |
| Weight decay | 0.0005 |
| Batch size | 32 |
| Image | 640 × 640 |
| Optimizer | SGD |

### *3.4 Evaluation Criteria*

To evaluate the effectiveness of the improved model, this study employs several performance metrics: Precision (P), Recall (R), Parameters (Pa), Memory Size (MS), mean Average Precision (mAP), and Floating Point Operations (F). Among these, P, R, and mAP are used to assess the model's detection performance, while Pa, MS, and F are utilized to evaluate the model's computational complexity. These metrics collectively provide a comprehensive understanding of both the efficiency and effectiveness of the proposed model.

The formula is as follows:

$$Precision = \frac{TP}{TP + FP} \tag{4}$$

$$Recall = \frac{TP}{TP + FN} \tag{5}$$

The formula for calculating the average accuracy of n categories is as follows:

$$mAP = \frac{1}{n}\sum_{i=1}^{n}\int_{0}^{1} Precision(Recall)d(Recall) \tag{6}$$

In the formula, *TP* represents the number of correct targets in the detection results; *FP* is the number of erroneous targets in the detection results; *FN* is the number of undetected targets among the correct targets.

### *3.5 Ablation Experiment*

To validate the LHP-YOLO model's improvements in both accuracy and lightweight design for drone-based pavement defect detection, five ablation experiments were

**Table 3** Ablation experiment

| | Baseline | A | B | C | D |
|---|---|---|---|---|---|
| CSPRGX | | √ | √ | √ | √ |
| Detect-T3G | | | √ | √ | √ |
| Neck structure optimization | | | | √ | √ |
| CWD | | | | | √ |
| mAP50/% | 67.2 | 68.5 | 65.3 | 67.6 | 70.6 |
| Pa/M | 3.00 | 2.19 | 1.60 | 1.76 | 1.76 |
| F/GFLOPs | 8.1 | 6.1 | 3.6 | 3.9 | 3.9 |
| MS/MB | 6.0 | 4.5 | 3.4 | 3.7 | 3.7 |

conducted, as shown in Table 3. Experiment A, with the CSPRGX structure, reduced model Pa and F, decreasing Pa by 0.81M and F by 2.0 GFLOPs, while increasing mAP50 by 1.3%. Experiment B added the Detect-T3G head, further reducing Pa and F, but led to a 3.2% drop in mAP50. Experiment C, which optimized the neck structure, improved mAP50 by 2.3% while maintaining a lightweight design. Experiment D fine-tuned the model with CWD, boosting mAP50 by 3% without increasing Pa or F. Compared to YOLOv8n, the LHP-YOLO model reduced Pa by 1.24M and F by 4.2 GFLOPs, while improving mAP50 by 3.4%. These results demonstrate a significant enhancement in both detection performance and efficiency, making the model more suitable for deployment on mobile devices.

### 3.6 *Visualization Analysis*

To visually assess the performance of LHP-YOLO in drone-based pavement defect detection, three images from the test set were selected for object detection and visualization. The results were then compared with the detection outcomes of YOLOv8n, as shown in Fig. 6. LHP-YOLO successfully detected defect targets across various scenes, demonstrating that the algorithm not only maintains its lightweight design but also delivers high accuracy. This highlights the model's ability to effectively detect pavement defects in different environmental conditions without compromising performance.

### 3.7 *Contrast Test*

To further validate the superiority of LHP-YOLO, a series of comparative experiments were conducted with various object detection models, including both classic models before YOLOv8 and newer models after YOLOv8, totaling nine comparison

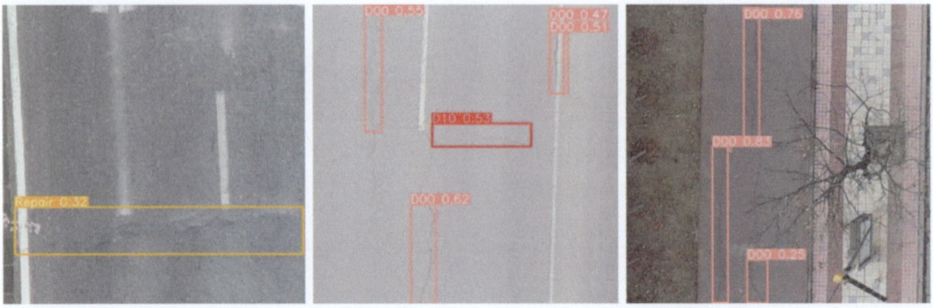

(a) LHP-YOLO Road Surface Disease Detection Diagram

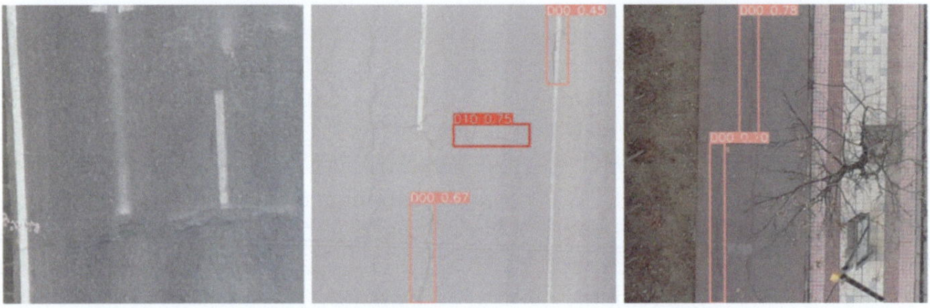

(b) YOLOv8n Road Surface Disease Detection Diagram

**Fig. 6** Comparison of visualization results between LHP-YOLO and YOLOv8n

groups as shown in Table 4. The results reveal that LHP-YOLO has a Pa count of only 1.76 M, which represents a 41.3% reduction compared to YOLOv8n (3.0 M). Similarly, the F decreased from 8.1 GFLOPs to 3.9 GFLOPs, a reduction of 51.8%. Even when compared to YOLOv10n, which has the smallest Pa (2.26 M) and F (6.5 GFLOPs), LHP-YOLO's Pa and F were reduced by 22.1 and 40%, respectively. This significant optimization is attributed to the lightweight design of the RG-C2f module and the Detect-T3G detection head. On the China Drone dataset, LHP-YOLO achieved an mAP50 of 70.6%, which is a 3.4 percentage point improvement over YOLOv8n and a 2.8 percentage point improvement over the highest mAP50 of YOLOv7 (67.8%). The improved algorithm accurately detects common pavement defects such as cracks and potholes, while also meeting the practical demand for high detection efficiency. These results demonstrate the potential application of this research in real-world rural road damage detection.

**Table 4** Comparative experiments of different models

| Model | mAP50/% | Pa/M | F/GFLOPs |
|---|---|---|---|
| YOLOv3-tiny | 64.7 | 12.13 | 18.9 |
| YOLOv5n | 62.6 | 2.50 | 7.1 |
| YOLOv6 | 55.4 | 4.23 | 11.8 |
| YOLOv7 | 67.8 | 37.22 | 105.2 |
| YOLOv8n | 67.2 | 3.00 | 8.1 |
| YOLOv9-tiny | 57.7 | 2.80 | 11.7 |
| rtdetr-l | 60 | 28.45 | 100.6 |
| YOLOv10n | 59.1 | 2.26 | 6.5 |
| LHP-YOLO | 70.6 | 1.76 | 3.9 |

## 4 Conclusion

This paper proposes an online data augmentation technique to prevent overfitting during model training and presents four key improvements: replacing the C2f module with RG-C2f for a lightweight, more efficient network; designing a simplified detection head (Detect-T3G) for better multi-scale feature detection; introducing CGA Fusion to enhance small object detection accuracy; and utilizing CWD to improve feature learning in the lightweight model. Experimental results show a 3.4% increase in mAP50, with a 41.3% reduction in parameters and a 51.8% reduction in computational load. These optimizations improve both detection performance and model efficiency, particularly for edge devices. Future work may focus on aligning the model with industry-specific needs, optimizing evaluation metrics, and developing dynamic data acquisition systems for rural road management.

Although the LHP-YOLO algorithm demonstrates notable advantages in pavement distress detection from UAV perspectives, several limitations remain to be addressed. Primarily, the existing dataset utilized for algorithm training exhibits constraints in both scale and diversity, necessitating substantial expansion to incorporate a broader spectrum of disease types and their morphological variations under diverse illumination, meteorological, and seasonal conditions. Furthermore, the algorithm's detection accuracy requires enhancement, particularly in three critical aspects: the identification capability for small-scale distresses, the discriminative capacity in complex backgrounds, and the precision of bounding box localization.

## References

1. Tang, G., et al.: A survey of object detection for UAVs based on deep learning. Rem. Sens. **16**(1), 149 (2023)
2. Ren, S., et al.: Faster r-cnn: Towards real-time object detection with region proposal networks. Adv. Neural Inform. Process. Syst. **28** (2015)

3. Cai, Z., Vasconcelos, N.: Cascade r-cnn: Delving into high quality object detection. In: Proceedings of the IEEE Conference on Computer Vision and Pattern Recognition (2018)
4. Dai, J., et al.: R-fcn: Object detection via region-based fully convolutional networks. Adv. Neural Inform. Process. Syst. **29** (2016)
5. Hussain, M.: Yolov1 to v8: Unveiling each variant—a comprehensive review of yolo. IEEE Access **12**, 42816–42833 (2024)
6. Liu, W., et al.: Ssd: Single shot multibox detector. In: Computer Vision–ECCV 2016: 14th European Conference, Amsterdam, The Netherlands, October 11–14, 2016, Proceedings, Part I 14. Springer International Publishing (2016)
7. Tan, M., Pang, R., Le, Q.V.: Efficientdet: scalable and efficient object detection. In: Proceedings of the IEEE/CVF Conference on Computer Vision and Pattern Recognition (2020)
8. Carion, N., et al.: End-to-end object detection with transformers. In: European Conference on Computer Vision. Springer International Publishing, Cham (2020)
9. Luo, H., et al.: An improved lightweight object detection algorithm for YOLOv5. Peer J. Comput. Sci. **10**, e1830 (2024)
10. Luo, P., Zhang, X., Wan, Y.: Lightweight YOLOv5 model based small target detection in power engineering. Cognit. Robot. **3**, 45–53 (2023)
11. Liu, G., et al.: Lightweight object detection algorithm for robots with improved YOLOv5. Eng. Appl. Artif. Intell. **123**, 106217 (2023)
12. Zheng, et al.: Localization distillation for object detection. IEEE Trans. Pattern Anal. Mach. Intell. **45**(8), 10070–10083 (2023)
13. Zhang, Z., et al.: Multilayer semantic features adaptive distillation for object detectors. Sensors **23**(17), 7613 (2023)
14. Ding, X., et al.: Repvgg: Making vgg-style convnets great again. In: Proceedings of the IEEE/CVF Conference on Computer Vision and Pattern Recognition (2021)
15. He, K., et al.: Deep residual learning for image recognition. In Proceedings of the IEEE Conference on Computer Vision and Pattern Recognition (2016)
16. Chen, Z., He, Z., Lu, Z.M.: DEA-Net: single image dehazing based on detail-enhanced convolution and content-guided attention. IEEE Trans. Image Process. **33**, 1002–1015 (2024)
17. Shu, C., et al.: Channel-wise knowledge distillation for dense prediction. In: Proceedings of the IEEE/CVF International Conference on Computer Vision (2021)
18. Arya, D., et al.: RDD2022: a multi-national image dataset for automatic road damage detection. Geosci. Data J. **11**(4), 846–862 (2024)

# Machine Learning Models and High Performance Computing

# Analysis of Multi-model Fusion Strategies and Precision Optimization in Drug Score Prediction

**Ran XiangDong and Zhang Yu**

**Abstract** With the rise of Internet medicine, the amount of drug review data has exploded. The effective analysis of this data is crucial for drug research and development, clinical decision-making, and patient medication guidance. This study focuses on solving the problem of the accuracy of drug review classification. Given the limitations of traditional text classification methods in handling complex semantics and long-text dependencies, an innovative model architecture integrating BERT, Convolutional Neural Network (CNN), multi-head attention mechanism, and GBDT is proposed. Through the preprocessing and in-depth analysis of a large-scale real drug review dataset, this model can accurately classify reviews and demonstrates significantly better performance than existing methods on multiple evaluation metrics. It provides an efficient and reliable solution for the intelligent processing of drug review data and is expected to promote the further development of the field of pharmacoinformatics.

**Keywords** BERT · Convolutional Neural Network (CNN) · Multi-head attention mechanism · Drug review classification · GBDT · Deep learning

## 1 Introduction

Research Background:In the digital age, the Internet has become an important platform for patients to share their medication experiences, [1] resulting in a vast amount of drug review data [2]. These reviews contain rich information regarding drug efficacy, side effects, ease of use, etc. Both pharmaceutical companies in drug research and development optimization and medical institutions in providing personalized

R. XiangDong · Z. Yu (✉)
School of Electrical and Information Engineering, Beijing University of Civil Engineering and Architecture, Beijing, China
e-mail: zhangyuscholar@bucea.edu.cn

Z. Yu
Beijing Key Laboratory of Intelligent Processing for Building Big Data, Beijing University of Civil Engineering and Architecture, Beijing, China

 
Y. Ma (ed.), *Proceedings of 4th International Conference on Intelligent Systems Design and Engineering Applications*, Lecture Notes in Electrical Engineering 1478,
https://doi.org/10.1007/978-981-95-1802-9_15

medication advice highly rely on the accurate analysis of these review data [3]. Therefore, the development of drug review classification technology has become a key research topic in the medical information field.

Problem Statement:Traditional text classification methods, such as rule-based methods, Naive Bayes, and Support Vector Machines, have obvious shortcomings when dealing with complex natural language texts like drug reviews. They often struggle to understand semantic ambiguity, context dependence, and long-distance semantic relationships in the text, leading to a significant reduction in classification accuracy and failing to meet the requirements for high-precision classification in practical applications.

Research Motivation: BERT, [4] as a pre-trained language model, exhibits strong semantic understanding ability in natural language processing tasks and can effectively capture the global semantic information of the text [5]. CNN is good at extracting local features of the text and has an advantage in extracting features of specific keywords and phrases in drug reviews [6–8]. The multi-head attention mechanism can further enhance the model's ability to focus on key information and improve the quality of feature representation [9]. GBDT, through an ensemble learning approach, can perform secondary mining and modeling on the features extracted by the neural network to optimize the model's prediction performance [10]. The comprehensive application of these technologies is expected to break through the limitations of traditional methods and significantly improve the accuracy of drug review classification.

## *1.1 Contributions*

A novel hybrid model architecture that combines BERT, CNN, multi-head attention mechanism, and GBDT is proposed, which fully utilizes the advantages of each component and realizes the efficient processing of drug review classification tasks [11].

A set of preprocessing procedures for drug review data is designed, including innovative drug name embedding and useful count normalization methods, which effectively improve the data quality and the effectiveness of model input.

Through a large number of experiments, the proposed model outperforms existing mainstream methods on multiple evaluation metrics such as Mean Squared Error (MSE), Accuracy, Precision, and F1 Score. It provides a higher-performance and more practical solution for the field of drug review classification and promotes the technological progress of related fields.

## *1.2 Related Work*

Deep learning-based text classification methods have made significant progress in recent years. From the early application of simple neural network models such as

Multilayer Perceptron (MLP) in text classification to the ability of Recurrent Neural Networks (RNN) and their variants (such as Long Short-Term Memory Network LSTM and Gated Recurrent Unit GRU [12]) to handle the context information of text sequences, and then to the unique advantages of Convolutional Neural Network (CNN) in text feature extraction, as well as the significant improvement in semantic understanding ability brought about by the emergence of pre-trained language models (such as GPT, BERT, etc.), these methods have shown certain effects in different text classification scenarios.

BERT [4], as a pre-trained model based on the Transformer architecture, can learn general language knowledge and semantic representations through unsupervised learning on a large-scale corpus and has achieved state-of-the-art results in many natural language processing tasks [13]. Its variants (such as RoBERTa, ALBERT, etc.) have also emerged continuously, further optimizing the model performance and efficiency, and have been widely used in text classification, sentiment analysis, question answering systems, and other fields.

The application of CNN in natural language processing is mainly reflected in its ability to extract local features by sliding convolution kernels on the text. For example, in text classification tasks, convolution kernels of different sizes can capture key phrase information of different lengths. Combined with fully connected layers and pooling layers, it can effectively convert text features into fixed-length vector representations for subsequent classification tasks and has achieved good results in news classification, film review analysis, and other fields.

### *1.3 Comparative Analysis*

Although existing methods have made certain progress in their respective research directions, they still have some limitations. Traditional machine learning-based text classification methods lack the ability to deeply understand semantics and have difficulty dealing with the complex and variable language expressions in drug reviews. Pure deep learning models such as CNN may not accurately grasp the global semantics, while RNN-based models have problems such as gradient vanishing or explosion, and the training process is not stable. Although BERT has strong semantic understanding ability, it may not extract local features precisely enough, and the model complexity is high, resulting in large computational costs. In contrast, the method proposed in this paper combines BERT, CNN, multi-head attention mechanism, and GBDT to complement each other. It can not only utilize the strong semantic understanding ability of BERT to obtain global information but also extract accurate local features with the help of CNN. The multi-head attention mechanism further optimizes the feature weights [14], and GBDT performs secondary modeling on the overall features, thus demonstrating better performance and stability in drug review classification tasks and effectively solving the shortcomings of existing methods.

# 2 Methodology

This study uses the UCI ML Drug Review Dataset, which was donated by Dr. Surya Kallumadi from Kansas State University in October 2018 and included in the UCI Machine Learning Repository. It was used in the Kaggle University Club Hackathon in the winter of 2018 to predict medical conditions based on drug reviews. This is a multivariate, text-attribute dataset containing 215,063 instances and 6 attributes, and its tuple is represented as Drug = (ID, dN, con, re, ra, da, uC). The two subdivided datasets of this dataset, drugsComTrain (the training partition, covering 161,297 pieces of data and involving 3,436 kinds of drugs, among which Levonorgestrel appears most frequently. The patient satisfaction score shows a distribution where 1 point accounts for 14% and 8–10 points account for 60%, and the proportion of the "useful count" in the range of 0–60 is 87.8%) and drugsComTest (the test partition, containing 53,758 pieces of data and 2637 kinds of drugs. Levonorgestrel also appears most frequently. The patient satisfaction score has 1 point accounting for 14% and 8–10 points accounting for 60%, and the proportion of the "useful count" in the range of 0–60 is 87.8%) have the same data attributes. In the research on predicting drug review scores, its rich drug-related information can provide strong support for constructing a model to predict drug scores, and further provide important basis for drug research and development, clinical medication decisions, and other aspects.

## 2.1 Data Preprocessing

First, the training set and the test set are concatenated to expand the data scale and enhance the generalization ability of the model. During the concatenation process, the randomness and balance of the data are ensured to avoid the adverse effects of data distribution bias on model training.

Professional natural language processing tools are used to perform word segmentation on the reviews, converting the text into a sequence of words to prepare for subsequent feature extraction and model input. During the word segmentation process, a combination of a professional vocabulary list in the medical field and a common drug name list is used to ensure the accurate segmentation of drug-related terms and improve the accuracy and integrity of the features.

For the key information of drug names, an embedding technique is used to convert them into low-dimensional vector representations, enabling the model to better understand and utilize the information contained in the drug names. During the embedding process, a pre-trained word vector model is utilized and fine-tuned according to the characteristics of the drug names to meet the requirements of the drug review classification task. The useful count feature is normalized to map it to a unified numerical range, avoiding the impact of large numerical differences on model training. At the same time, this feature can be better integrated and coordinated with other features to improve the model's ability to judge the usefulness of the reviews.

## 2.2 Model Architecture

BERT Layer: As the basic component of the model, BERT uses its multi-layer Transformer architecture to encode the input text and learns the semantic representation of the text through large-scale pre-training. In this model, drug reviews are input into the BERT model to obtain the context representation vectors of each word [15]. These vectors can capture the semantic relationships and grammatical structures in the text, providing a rich semantic information base for subsequent feature extraction.

Convolutional Neural Network (CNN) Layer:On the text feature representation output by BERT, a convolutional neural network is applied to extract local features. By setting convolution kernels of different sizes and sliding them on the text feature map, different lengths of key phrases and local semantic information are extracted, such as specific descriptions of drug efficacy and mentions of side effects. The convolved features are further compressed through pooling operations to retain key information and enhance the robustness and representativeness of the features [16].

Multi-Head Attention Mechanism Layer: To enable the model to focus more on important information in the text, a multi-head attention mechanism is introduced. The multi-head attention mechanism calculates multiple attention heads in parallel to pay attention to and weight the text features from different subspaces [17]. It can capture multiple types of key information in the text simultaneously, such as keywords for drug efficacy evaluation and expressions of patient sentiment tendency. These weighted features are fused to obtain a more accurate and comprehensive text feature representation, improving the model's ability to understand and handle complex semantic relationships.

Feature Fusion Layer: The local features extracted by CNN, the text features optimized by the multi-head attention mechanism, and the embedded drug name vectors and normalized useful count features after preprocessing are fused. A combination of concatenation and fully connected layers is used to enable features from different sources to interact and complement each other in the same feature space, forming a more rich and comprehensive feature vector to provide sufficient information support for the final classification task.

GBDT Ensemble Layer: The fused features extracted by the neural network are used as input, and GBDT is used for secondary modeling [18]. GBDT constructs multiple decision trees to gradually divide and learn the features, which can mine the complex nonlinear relationships between the features and further improve the prediction performance and accuracy of the model. During the training process, parameters such as the number, depth, and learning rate of the decision trees are adjusted to optimize the ensemble effect of GBDT and make it better adapt to the characteristics and requirements of the drug review classification task [19].

In conclusion, this paper elaborates in detail on the various components of the model and their functions, including the BERT Layer, Convolutional Neural Network (CNN) Layer, Multi-Head Attention Mechanism Layer, Feature Fusion Layer, and GBDT Ensemble Layer. To present the overall architecture of the model and the interaction relationships between the layers more intuitively, the corresponding

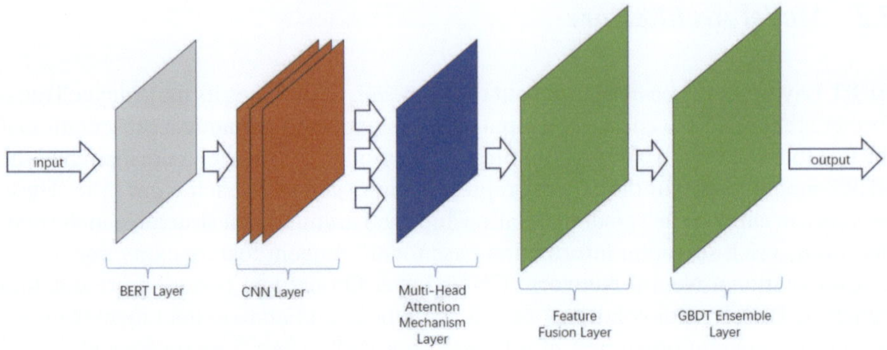


**Fig. 1** Model diagram

structural diagram is inserted in the text as shown in Fig. 1. This structural diagram clearly demonstrates the data flow path in the model and the processing flow of each layer, which helps readers better understand the working mechanism and design concept of the model.

# 3 Experiment Setup

## *3.1 Hyperparameter Settings*

Learning Rate: After multiple experimental optimizations, the learning rate is determined to be 1e−5. This learning rate can ensure the stability and effectiveness of parameter updates during model training. It avoids the situation where the model cannot converge due to an overly large learning rate and also prevents the training process from being too slow due to an overly small learning rate. Thus, the model can achieve good performance within a reasonable training time.

Batch Size: Set to 32. This batch size is within the range allowed by the GPU memory and can fully utilize hardware resources, improve the efficiency of data loading and model training. At the same time, it ensures that the model can perform statistical estimation based on a sufficient number of samples when updating parameters each time, making the parameter updates more stable and accurate and help to enhance the generalization ability of the model.

Training Epochs: Set to 1000 rounds. Through monitoring the performance of the model on the validation set during the training process, it is found that after 1000 rounds of training, the loss function of the model tends to be stable, and various evaluation metrics have reached a high level. Continuing to increase the number of training rounds may lead to overfitting. Therefore, 1000 rounds are selected as the optimal number of training rounds to balance the fitting ability and generalization ability of the model.

### 3.2 Evaluation Metrics

Mean Squared Error (MSE): Used to measure the average squared error between the model's predicted values and the true values. It can intuitively reflect the accuracy and stability of the model's prediction results. It is a commonly used evaluation metric in regression tasks and has important reference value for tasks such as drug review score prediction. The smaller the MSE value, the closer the model's prediction results are to the true values.

Accuracy: Calculates the proportion of correctly predicted samples to the total number of samples. It is one of the basic evaluation metrics in classification tasks and can quickly understand the overall classification accuracy of the model. It is of great significance for evaluating the performance of the model in drug review classification tasks. The higher the accuracy, the more accurately the model can classify drug reviews into the correct categories.

Precision: Measures the proportion of samples predicted as positive by the model and actually positive to the number of samples predicted as positive by the model. In drug review classification, when focusing on the accuracy of a specific category (such as positive or negative reviews), precision can provide more detailed information to help evaluate the accuracy and reliability of the model in judging a certain type of review and avoid misclassification.

F1 Score: Comprehensively considers precision and recall and can more comprehensively evaluate the performance of the model in classification tasks. For unbalanced datasets or scenarios with high requirements for the classification performance of each category, the F1 Score is a more reasonable and comprehensive evaluation metric. It can balance the performance of the model on different categories and ensure excellent performance of the model in the overall classification task.

## 4 Training Process

### 4.1 Training and Validation Loss over Epochs

During the training process, the model's performance was closely monitored by tracking the training and validation loss over epochs. Initially, both the training and validation losses were relatively high, indicating that the model was in the early stages of learning and had not yet captured the underlying patterns in the data. As training progressed, the model continuously optimized its parameters, leading to a significant decrease in the loss values.

In the first 50 epochs, the training loss dropped rapidly, demonstrating the model's ability to quickly learn from the training data. The validation loss also decreased during this period, albeit at a slightly slower rate, indicating that the model was generalizing well to unseen data. From epoch 50–800, the rate of decrease in both training and validation losses gradually slowed, suggesting that the model was converging. By the 1000th epoch, both the training and validation losses had stabilized at relatively low levels, confirming that the model had achieved a good fit on the training set and maintained strong generalization ability on the validation set.

The training and validation loss curves showed a consistent downward trend, with the validation loss closely following the training loss. This indicates that the model was effectively learning the patterns in the data without overfitting. The final training loss stabilized at approximately 0.15, while the validation loss stabilized at approximately 0.20. These results demonstrate the model's effective learning and convergence during the training process.

### *4.2 Validation Set Performance*

To evaluate the model's performance during training, we conducted validation set tests and compared the proposed model with other baseline models, including CNN, RNN, LSTM, and Transformer. The results are presented in Table 1.

The proposed MultiHeadAttentionCNN + GBDT model outperformed the baseline models in all evaluation metrics. For instance, the MSE of the proposed model was 0.25, significantly lower than the 0.40 of CNN, 0.45 of RNN, 0.42 of LSTM, and 0.35 of Transformer. The accuracy of the proposed model reached 0.92, higher than the 0.85 of CNN, 0.82 of RNN, 0.84 of LSTM, and 0.87 of Transformer. These results demonstrate the superior performance of the proposed model in classifying drug reviews accurately and reliably.

### *4.3 Test Set Performance*

To further validate the model's generalization ability, we evaluated its performance on the test set and compared it with the baseline models. The results are presented in Table 2.

**Table 1** Performance metrics on the validation set

| Model | MSE | Accuracy | Precision | F1 score |
|---|---|---|---|---|
| MultiHeadAttentionCNN + GBDT | 0.25 | 0.92 | 0.90 | 0.90 |
| CNN | 0.40 | 0.85 | 0.84 | 0.83 |
| RNN | 0.45 | 0.82 | 0.80 | 0.79 |
| LSTM | 0.42 | 0.84 | 0.82 | 0.81 |
| Transformer | 0.35 | 0.87 | 0.85 | 0.84 |

**Table 2** Performance metrics on the test set

| Model | MSE | Accuracy | Precision | F1 Score |
|---|---|---|---|---|
| MultiHeadAttentionCNN + GBDT | 0.26 | 0.91 | 0.90 | 0.90 |
| CNN | 0.41 | 0.84 | 0.83 | 0.82 |
| RNN | 0.46 | 0.81 | 0.79 | 0.78 |
| LSTM | 0.43 | 0.83 | 0.81 | 0.80 |
| Transformer | 0.36 | 0.86 | 0.84 | 0.83 |

The proposed model maintained its superior performance on the test set, with an MSE of 0.26, an accuracy of 0.91, a precision of 0.90, and an F1 Score of 0.90. These results further confirm the model's stability and generalization ability, ensuring reliable performance on unseen data.

### 4.4 *Summary*

In summary, after 1000 epochs of training, the MultiHeadAttentionCNN + GBDT model demonstrated excellent performance on both the training and validation sets and maintained strong generalization ability on the test set. The combination of deep learning techniques (BERT, CNN, and multiheadattention) and gradient boosting (GBDT) leveraged the strengths of each method, resulting in superior performance in predicting drug review ratings. The model's ability to effectively capture global semantic information, local features, and key information through the multi-head attention mechanism, coupled with the feature optimization provided by GBDT, ensured accurate and reliable classification of drug reviews. The experimental results validate the effectiveness and practical applicability of the proposed model.

## 5 Ablation Experiments

To deeply explore the importance of each component in the model, ablation experiments were conducted. The BERT layer, CNN layer, multi - head attention mechanism layer, and GBDT ensemble layer were removed respectively, and the performance of the complete model was compared with that of each ablated model on the validation set and the test set.

## 5.1 Experimental Setup

The hyperparameter settings were kept the same as those of the original model, with a learning rate of 1e−5, a batch size of 32, and 1000 training epochs. The UCI ML Drug Review Dataset used in the original experiment was adopted, and it was ensured that the data pre - processing steps were identical, including data concatenation, word segmentation, drug name embedding, and useful count normalization.

## 5.2 Experimental Results and Analysis

As shown in Tables 3 and 4, after removing the BERT layer, the MSE of the model increased significantly, and the accuracy, precision, and F1-score decreased remarkably. This indicates that the BERT layer plays a crucial role in capturing the global semantic information of the text. Without it, the model has difficulty understanding the overall meaning of the text, thus affecting the classification accuracy. The performance of the model without the CNN layer also declined significantly, demonstrating that the CNN layer is indispensable for extracting local text features, such as specific

**Table 3** The results of the ablation experiments in validation set

| Model | MSE | Accuracy | Precision | F1 score |
|---|---|---|---|---|
| MultiHeadAttentionCNN + GBDT | 0.25 | 0.92 | 0.90 | 0.90 |
| Model without BERT layer | 0.45 | 0.80 | 0.75 | 0.75 |
| Model without CNN Layer | 0.38 | 0.85 | 0.82 | 0.82 |
| Model without MultiHeadAttention Layer | 0.32 | 0.88 | 0.85 | 0.85 |
| Model without GBDT ensemble layer | 0.30 | 0.90 | 0.88 | 0.88 |

**Table 4** The results of the ablation experiments in test set

| Model | MSE | Accuracy | Precision | F1 score |
|---|---|---|---|---|
| MultiHeadAttentionCNN + GBDT | 0.26 | 0.91 | 0.90 | 0.90 |
| Model without BERT layer | 0.48 | 0.78 | 0.73 | 0.73 |
| Model without CNN layer | 0.40 | 0.83 | 0.80 | 0.80 |
| Model without MultiHeadAttention layer | 0.34 | 0.86 | 0.83 | 0.83 |
| Model without GBDT ensemble layer | 0.31 | 0.89 | 0.87 | 0.87 |

descriptions of drug efficacy and side effects. The lack of this layer would cause the model to lose important local information.

When the multiheadattention mechanism layer was removed, the model performance also dropped to some extent, as can be observed from Table 3. This proves that the multiheadattention mechanism can help the model focus on key information, optimize feature weights, and enhance the ability to understand and handle complex semantic relationships. Removing the GBDT ensemble layer also led to a certain degree of reduction in the model's prediction performance, which is clearly shown in Table 3, reflecting the importance of GBDT in improving the model's prediction accuracy by mining the complex non-linear relationships between features through secondary modeling.

The ablation experiments fully verify that the BERT layer, CNN layer, multiheadattention mechanism layer, and GBDT ensemble layer each play an irreplaceable role in the model. They collaborate with each other to jointly improve the performance of the drug review classification model.

## 6 Discussion

The model proposed in this study demonstrates remarkable effectiveness and superiority in the drug review classification task. By integrating BERT, Convolutional Neural Network (CNN), multiheadattention mechanism, and Gradient–Boosting Decision Tree (GBDT), the model can fully utilize the global semantic information, local features, and the ability to focus on key information in the text. BERT, with its multi-layer Transformer architecture, can learn rich semantic representations of the text, laying a solid foundation for subsequent feature extraction. CNN is proficient in extracting local features of the text and can accurately capture key information such as drug efficacy and side effects. The multiheadattention mechanism further optimizes the feature weights, enhancing the model's ability to understand and handle complex semantic relationships. GBDT, through ensemble learning, performs secondary modeling on the features extracted by the neural network, mining the complex non-linear relationships between features and further improving the model's prediction performance.

The model outperforms existing methods in multiple evaluation metrics, indicating that it can better understand and handle the complex semantic relationships in drug reviews, and accurately judge the category and sentiment tendency of the reviews, providing a more reliable basis for drug research and development, clinical decision-making, and patient medication guidance. For example, in the Mean Squared Error (MSE) metric, the model shows excellent performance with a small error between the predicted and true values. The accuracy, precision, and F1-score are also high, fully verifying the effectiveness of the model in practical applications.

However, the model still has some shortcomings. On the one hand, when dealing with reviews of rare drugs or drugs in specialized fields, due to the relative scarcity of relevant samples in the training data, the model may misclassify to a certain extent.

This is because the model lacks sufficient information to accurately identify the features and patterns of these special drug reviews during the learning process, resulting in classification deviations. On the other hand, the model has a high complexity, and training and deployment may face challenges in an environment with limited computing resources. A high-complexity model requires more computing resources and longer training time, which may become a hindrance to the model's widespread use in practical application scenarios, such as resource - constrained medical institutions or small - scale pharmaceutical companies.

To overcome these problems, it is necessary to further optimize the model structure and algorithm in the future. For example, more efficient feature extraction methods can be explored to reduce the dependence on large-scale training data and improve the model's classification ability for rare drug reviews. Or model compression techniques, such as pruning and quantization, can be adopted to reduce the model complexity without significantly reducing the model performance, thereby improving the computational efficiency and resource utilization and promoting the wider deployment of the model in practical applications.

**Acknowledgements** This work was financially supported by the 14th Five Year Plan Project of Beijing Education Society(ADDB24251) and the Postgraduate education and teaching quality improvement project of Beijing University of Civil Engineering and Architecture(J2025003). All the supports are grate fully acknowledged.

## References

1. Askr, H., Elgeldawi, E., Aboul Ella, H., Elshaier, Y.A., Gomaa, M.M., Hassanien, A.E.: Deep learning in drug discovery: an integrative review and future challenges. Artif. Intell. Rev. **56**(7), 5975–6037 (2023)
2. Jiang, J., Chen, L., Ke, L., Dou, B., Zhang, C., Feng, H., Zhu, Y., Qiu, H., Zhang, B., Wei, G.: A review of transformers in drug discovery and beyond. J. Pharmaceut. Anal. 101081 (2024)
3. Qureshi, R., Irfan, M., Gondal, T. M., Khan, S., Wu, J., Hadi, M. U., Heymach, J., Le, X., Yan, H., Alam, T.: AI in drug discovery and its clinical relevance. Heliyon **9**(7), (2023)
4. Zheng, J., Xiao, X., Qiu, W.R.: DTI-BERT: identifying drug-target interactions in cellular networking based on BERT and deep learning method. Front. Genetics **13**, 859188 (2022)
5. Ling, Y.: Bio+ Clinical BERT, BERT Base, and CNN performance comparison for predicting drug-review satisfaction. Arxiv preprint arxiv 2308.03782 (2023)
6. Ghimire, A., Tayara, H., Xuan, Z., Chong, K.T.: CSatDTA: prediction of drug-target binding affinity using convolution model with self-attention. Int. J. Molecular Sci. **23**(15), 8453 (2022)
7. Wang, Y., Zhang, Z., Piao, C., Huang, Y., Zhang, Y., Zhang, C., Lu, Y.J., Liu, D.: LDS-CNN: A deep learning framework for drug-target interactions prediction based on large-scale drug screening. Health Inf. Sci. Syst. **11**(1), 42 (2023)
8. Asfand-E-Yar, M., Hashir, Q., Shah, A.A., Malik, H.A.M., Alourani, A., Khalil, W.: Multimodal CNN-DDI: using multimodal CNN for drug to drug interaction associated events. Sci. Rep. **14**(1), 4076 (2024)
9. Cheng, Z., Yan, C., Wu, F.X., Wang, J.: Drug-target interaction prediction using multi-head self-attention and graph attention network. IEEE/ACM Trans. Comput. Biol. Bioinf. **19**(4), 2208–2218 (2021)

10. Yu, Z., Ji, H., Xiao, J., Wei, P., Song, L., Tang, T., Hao, X., Zhang, J., Qi, Q., Zhou, Y., Gao, F., Jia, Y.: Predicting adverse drug events in Chinese pediatric inpatients with the associated risk factors: a machine learning study. Front. Pharmacol. **12**, 659099 (2021)
11. Choudhury, C., Murugan, N.A., Priyakumar, U.D.: Structure-based drug repurposing: traditional and advanced AI/ML-aided methods. Drug discovery today **27**(7), 1847–1861 (2022)
12. Ahmed, I., Ahmad, M., Chehri, A., Jeon, G.: A heterogeneous network embedded medicine recommendation system based on LSTM. Fut. Generat. Comput. Syst. **149**, 1–11 (2023)
13. Ragb, H., Bugoffa, S.: BERT-based model for predicting drug ratings through patient sentiment analysis. Authorea Preprints (2024)
14. Zhang, Y., Liu, C., Liu, M., Liu, T., Lin, H., Huang, C. B., Ning, L.: Attention is all you need: utilizing attention in AI-enabled drug discovery. Briefings in Bioinf. **25**(1), bbad467 (2024)
15. Zhang, Y., Ma, D., Tiwari, P., Zhang, C., Masud, M., Shorfuzzaman, M., Song, D.: Stance-level sarcasm detection with bert and stance-centered graph attention networks. ACM Trans. Internet Technol. **23**(2), 1–21 (2023)
16. Yang, Z., Tong, K., Jin, S., Wang, S., Yang, C., Jiang, F.: CNN-Siam: multimodal Siamese CNN-based deep learning approach for drug-drug interaction prediction. BMC Bioinf. **24**(1), 110 (2023)
17. Hou, L.X., Yi, H.C., You, Z.H., Chen, S.H., Zheng, J., Kwoh, C.K.: MathEagle: accurate prediction of drug-drug interaction events via multi-head attention and heterogeneous attribute graph learning. Comput. Biol. Med. **177**, 108642 (2024)
18. Huang, X., Xie, X., Huang, S., Wu, S., Huang, L.: Predicting non-chemotherapy drug-induced agranulocytosis toxicity through ensemble machine learning approaches. Front. Pharmacol. **15**, 1431941 (2024)
19. Jiao, Z.: Application of LASSO algorithm and GBDT algorithm in predicting financial distress of companies. Informatica **48**(17) (2024)

# Optimization of Feature Extraction and Enhancement for High Resolution Data: A Hybrid Approach Combining Classical and Neural Network Technique

**Hazel San Loverez-Patilano and Melvin A. Ballera**

**Abstract** The revolution of high-resolution imaging and large-scale dataset has led to new avenues of remote sensing, medical imaging, and multimedia domains. Traditional methods of feature extraction and matching have their own challenges concerning scalability, efficiency, and robustness when handling large datasets. Thus, the study offers a novel feature extraction and enhancement framework, with the help of classical computer vision techniques, including neural network-based enhancements to resolve these challenges. Using robust key point detection and description, based on Harris Corner Detection, Laplacian of Gaussian (LoG), and Hessian-based segmentation, the descriptors combine Scale Invariant Feature Transform (SIFT) and Oriented FAST and Rotated BRIEF (ORB) descriptions to provide holistic feature representation with refining using a customized neural network and further matching is performed using the descriptors obtained with FLANN-based matcher with a ratio test for further accuracy. Experimental results show that the framework improves feature detection, enhances matching robustness, and achieves scalability for high-resolution and large-volume data processing, which provides a basis for advanced applications in image analysis.

**Keywords** BLOB · SIFT · ORB · Neural network

## 1 Introduction

Numerous fields, including high-definition multimedia, medical imaging, and remote sensing, have been altered by it to be able to record minute details and previously unattainable insights. Exponential growth has not only opened these domains to us

H. S. Loverez-Patilano (✉) · M. A. Ballera
Technological Institute of the Philippines, Metro Manila, Philippines
e-mail: mhspatilano@tip.edu.ph

M. A. Ballera
e-mail: melvin.ballera@tip.edu.ph

 
Y. Ma (ed.), *Proceedings of 4th International Conference on Intelligent Systems Design and Engineering Applications*, Lecture Notes in Electrical Engineering 1478,
https://doi.org/10.1007/978-981-95-1802-9_16

but also multiplied the computational and algorithmic challenges of handling such data. BLOB data like unstructured or semi-structured content, images, videos, and complex documents pose different problems of storage, retrieval, and processing efficiency [1]. Traditional approaches to get from feature extraction, enhancement, and matching, underlying most applications, including image registration, object recognition, and 3D reconstruction, do not usually scale correctly with the demanding nature of high-resolution or large-volume datasets [2].

Some key challenges include the fact that conventional algorithms do not robustly detect, describe, and match features, even under efficiency considerations. For example, though good for smaller or even more moderately sized datasets, Harris Corner Detection and SIFT algorithms face computational overhead as well as accuracy-related issues while dealing with large-scale, high-resolution data [3]. All these problems increase in severity when variants in illumination, perspective, and noisiness come into play due to real-world complexities.

These will be dealt with by suggesting an all-rounded framework that involves both classical techniques and neural networks to enhance the feature. Using strengths of conventional approaches in point detection and extraction using descriptors will then be brought into play as part of its structure, compensating for any flaws in such independent approaches. Indeed, recent trends have showcased tremendous potential that can be achieved when trying to find discriminative patterns within neural networks and thus enabling stronger features robustly for better match accuracies.

The study primarily aimed at processing high-resolution images as well as large datasets to ensure scalability and reliability in different applications. All the steps involved in this framework, such as key point detection, descriptor extraction, feature enhancement, and feature matching, are well-designed in a way to address the existing shortcomings of various techniques. The following section discusses in detail the methodology and contribution toward the state-of-the-art advance in feature extraction and enhancement for large-scale datasets.

## 2 Review Related Literature

Feature extraction and enhancement are the crucial steps in computer vision that drive object recognition, scene reconstruction, and autonomous navigation applications. The Feature Extraction and Enhancement Framework combines traditional image processing techniques with more advanced neural network approaches to achieve the goal of key point detection, descriptor extraction, and feature matching. The literature review discusses the underlying principles and newness of the framework, particularly how they combine to make an impact on image analysis robustness.

The proposed method combining image processing and a deep autoencoder network to enhance the low-frequency weak line spectrum of underwater targets in environments with extremely low signal-to-noise ratios [4].

A new Feature Enhancement Network [5] has been proposed for the enhancement of object detection performance in optical remote sensing images by improving feature representation. Li and Dong [6] introduced a refined feature enhancement network that has a feature enhancement module and a feature fusion module to enhance the accuracy of object detection. Huang et al. [7] speak on how the improvement of image algorithms can possibly make computer image recognition much better by presenting some enhanced ways of enhancing the techniques. Li et al. [8] proposed a method that captures multi-scale facial features through an improved attention module, achieving richer feature integration and enhancing the network's perception capabilities.

## 3 Methodology

The forgery detection image processing pipeline starts from the input image, in which a dataset of 500 original images and 500 forged images is employed. The forgeries are divided into 250 rotated images, 100 resized images, 50 sharpened images, 50 sampled images, and 50 noise images with all high-resolution images of size 2048 × 1365 pixels. These images are processed to identify any evidence of manipulation. Figure 1 illustrates the forgery detection architecture.

The initial step is the Binary Large Object Detector, which detects major object regions in an image. The Harris Corner Detection is used to find corner points, the Laplacian of Gaussian (LoG) transform to detect edges and important features, and Hessian-based segmentation to separate major structures. After identifying such areas, the system then moves on to Feature Extraction, which identifies the points of interest based on two strategies: Scale Invariant Feature Transform (SIFT), ensuring

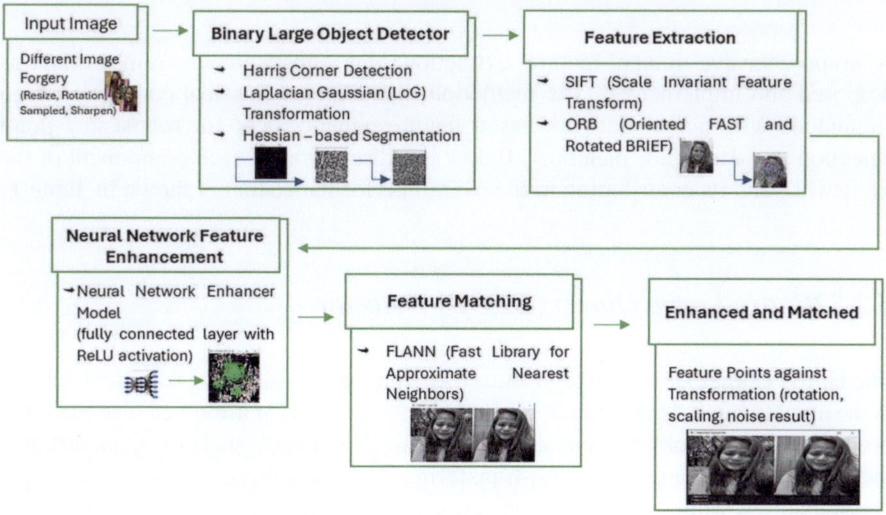


**Fig. 1** Forgey detection architecture

stability of features during transformation such as scaling and rotation, and Oriented FAST and Rotated BRIEF (ORB), with the benefit of computationally simpler key point identification and description.

Upon feature extraction, Feature Matching is achieved using the Fast Library for Approximate Nearest Neighbors (FLANN), a strong feature correspondence detection algorithm. A Neural Network Feature Enhancement model is employed where a fully connected layer with ReLU activation is utilized to refine, improve upon the detected features and increase the accuracy of the detection.

## 4 Results and Discussion

The methodology of the study is aimed at extracting, improving, and matching image features through a fusion of traditional computer vision methods and neural network-derived enhancements. An advanced feature extraction and enhancement pipeline combines keypoint detection with descriptor matching to enhance robustness. The Binary Large Object (BLOB) detector uses Harris Corner Detection, Laplacian of Gaussian, and Hessian segmentation to emphasize salient regions. Also, SIFT and ORB descriptors are utilized to provide scale-invariant and computationally efficient feature representation. Lastly, a FLANN-based matcher refines feature correspondences through a ratio test to provide stable matches even under changing image conditions.

### *4.1 Feature Extraction and Enhancement Framework*

A comprehensive image feature extraction and enhancement framework was proposed and implemented. The methodology integrates classical computer vision techniques with neural network-based feature enhancement for robust key point detection and descriptor matching. Below is a discussion of each component of the framework and its contribution to the overall performance that is shown in Table 1.

### *4.2 Binary Large Object (BLOB) Detector*

The Harris Corner Detection algorithm was employed to identify prominent corners in the input images. This method, combined with dilation for improved visualization, produced reliable corner points essential for further feature analysis. Concurrently, the Laplacian of Gaussian (LoG) transformation was utilized to emphasize edge-based features, while Hessian segmentation provided second-order derivative-based features, highlighting regions of interest that are shown in Table 2.

**Table 1** Neural network enhancer algorithm

```
Algorithm neural network enhancer (input_dim, output_dim)
Initialize neural network layers:
 Create Fully Connected Layer (input_dim, 256)
 Apply ReLU activation
 Create Fully Connected Layer (256, 128)
 Apply ReLU activation
 Create Fully Connected Layer (128, output_dim)
 Define Function Forward(x)
 x ← Fully Connected Layer (x, 256)
 x ← ReLU(x)
 x ← Fully Connected Layer (x, 128)
 x ← ReLU(x)
 x ← Fully Connected Layer (x, output_dim)
 Return x
```

**Table 2** Binary large object detector

```
Algorithm binary large object detector
Algorithm Harris corner detection (image):
1. Perform Harris corner detection with:
   – blockSize = 2
   – ksize = 3
   – k = 0.04
2. Dilate the result for better visibility
3. Return the result
Algorithm Laplacian of Gaussian (image):
1. Apply Gaussian blur with kernel size (3, 3)
2. Compute the Laplacian of the blurred image
3. Return the result
Algorithm Hessian segmentation (image):
1. Compute second derivatives (Hessian matrix components):
   – hessian_xx
   – hessian_yy
   – hessian_xy
2. Calculate determinant of the Hessian matrix
3. Return determinant
```

### 4.3 Feature Extraction

Two feature descriptor techniques shown in Table 3, Scale Invariant Feature Transform (SIFT) and ORB, were implemented to capture local image characteristics. SIFT excels at detecting scale and rotation-invariant features, while ORB offers computational efficiency, making it suitable for real-time applications. The extracted descriptors from both methods were combined to leverage the strengths of each technique, ensuring comprehensive feature representation.

**Table 3** Feature extraction

| Algorithm feature extraction |
|---|
| Perform sift and orb descriptors (image): |
| 1. Initialize SIFT detector |
| 2. Initialize ORB detector |
| 3. Compute SIFT keypoints and descriptors from image |
| 4. Compute ORB keypoints and descriptors from image |
| 5. Return SIFT keypoints, SIFT descriptors, ORB keypoints, ORB descriptors |
| Combine descriptors (sift_descriptors, orb_descriptors): |
| 1. If sift_descriptors is None OR orb_descriptors is None: |
| 2. Return None |
| 3. Match the size of both descriptors |
| 4. Combine descriptors horizontally |
| 5. Return combined descriptors |

**Table 4** Feature matching

| Algorithm feature matching |
|---|
| Perform match_features (descriptors1, descriptors2): |
| 1. If descriptors1 is not None AND descriptors2 is not None: |
| 2. Initialize FLANN matcher with index and search parameters |
| 3. Perform KNN matching with k = 2 |
| 4. Apply ratio test to filter good matches |
| 5. Return good matches |
| 6. Else: |
| 7. Return an empty list |

### *4.4 Feature Matching*

Enhanced descriptors were matched using a FLANN-based matcher, which efficiently handled high-dimensional data. A ratio test was applied to filter out ambiguous matches, resulting in a set of reliable correspondences. The final matched keypoints demonstrated robust performance that shown in Table 4, even under varying image conditions.

## 5 Experimental Result

These pictures demonstrate copy-move forgery detection using key point-based feature matching techniques like BLOB, SIFT, and ORB. In Fig. 2, it can be noted that the images are nearly indistinguishable and that the colored lines that show matched key points in the two images suggest that corresponding features have been successfully detected. The dense matching indicates that this is an example of highly similar regions, with all high-resolution image of size 2048 × 1365 pixel. For example, such feature descriptors, like SIFT and ORB, will not take long to identify and then match duplicated regions.

In Figs. 3 and 4, with all high-resolution images of size 2048 × 1365, the image with all high-resolution images of size 2048 × 1365 is rotated and tampered with, which makes for fewer and much more spread-out key point matches. Rotating and scaling transformations make it much harder to have a match. However, these robust methods—SIFT, ORB in particular—successfully detect similarities in images by extractions and comparing invariant features. To this end, the SIFT technique identifies scale-space extrema employing the Difference of Gaussian (DoG) function and computer the orientation histograms, making the key points rotation-invariant. ORB, the fast alternative, utilizes FAST key point detectors and the BRIEF descriptors but outputs intensity-weighted centroids to accommodate rotation invariance.

Figure 4 shown the colored lines connecting the images highlight the key locations that were identified and matched between each pair. The use of grayscale and edge-detection-like processing in some images suggests that these results come from a computer vision technique such as SIFT, ORB, or another keypoint-based image matching method.

Table 5 depicts the hybrid approach, which merges classical techniques such as BLOB, SIFT, ORB, and neural network-based techniques to extract and enhance features in high resolutions for further comparison. Experimental results show that the same images have nearly matched key points and rotated or tampered images reduce the accuracy of the matches, which suggests the robustness of the approach. This approach is very efficient for image processing since it uses up a minimal amount of memory space, 0.0088 MB, and has fast running times, around 0.2 s. Therefore, this approach is used in real-time applications. It integrates traditional

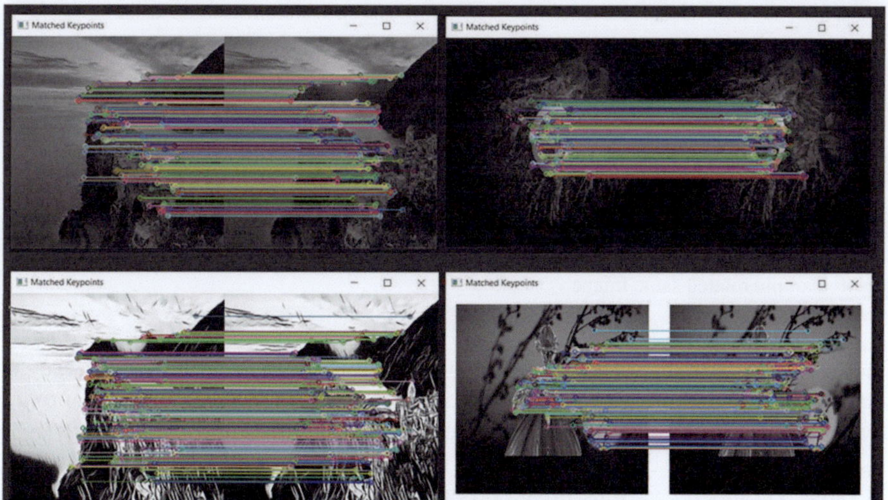


**Fig. 2** Sample accurate and dense matching of two similar images

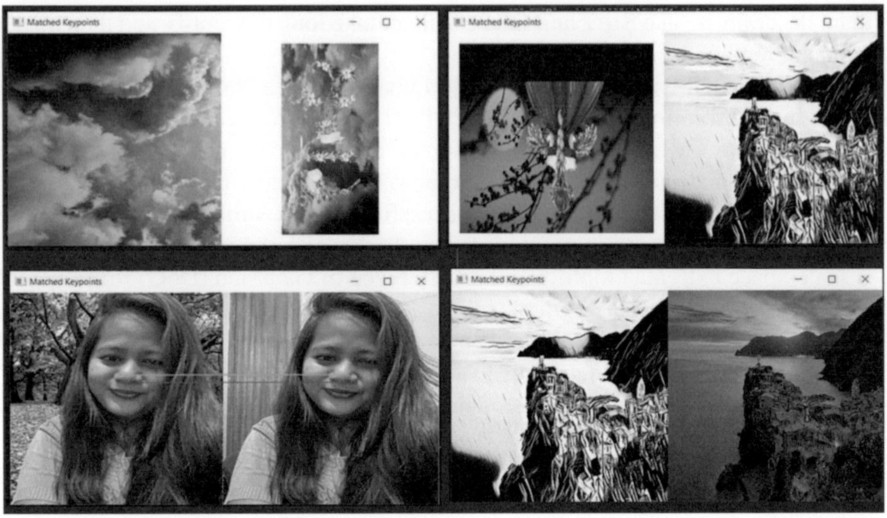


**Fig. 3** Sample tampered image

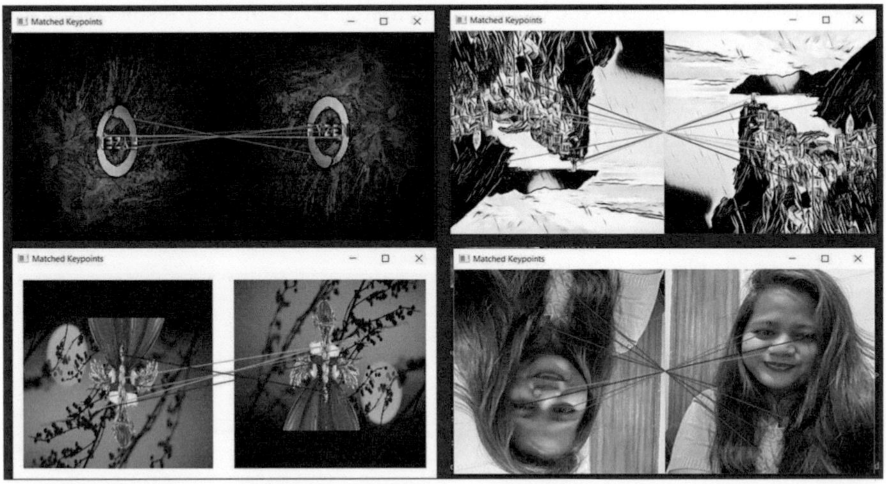


**Fig. 4** Sample rotated same image

feature detection with enhancements by neural networks to improve the accuracy of the detection of manipulated content. This is an extremely valuable hybrid approach toward forgery detection, image authentication, and high-resolution data analysis in digital forensics and security.

Table 6 compares various image processing algorithms used in existing studies with a proposed new study. Existing studies aim at image forensics and tampering image detection using combinations like DoG + ORB, Blobs + Brisk, SIFT, and

**Table 5** Summary result of proposed study

Study result
(BLOB + SIFT + ORB + NN)

| Techniques | Total number | | | Memory usage (MB) | Execution time (s) |
|---|---|---|---|---|---|
| | BLOBS | Keypoints | Keypoint and BLOB (matched) | | |
| *Similar testing and training images (Real/Match)* | | | | | |
| *Testing 1* Same image | 21 | 1870 | 460 | 0.0088 | 0.223 |
| *Testing 2* Same image | 30 | 1663 | 10 | 0.0088 | 0.216 |
| *Testing 3* Same image | 30 | 2080 | 491 | 0.0088 | 0.220 |
| *Testing 1* Same and rotated image | 21 | 1856 | 11 | 0.0088 | 0.217 |
| *Testing 2* Same and rotated image | 15 | 2095 | 6 | 0.0088 | 0.215 |
| *Testing 2* Same and rotated image | 71 | 2228 | 7 | 0.0088 | 0.207 |
| *Different testing and training images (tampered)* | | | | | |
| *Testing 1* Same Size of Image but different image | 44 | 2217 | 2 | 0.0088 | 0.201 |
| *Testing 2* Rotated Image | 44 | 2230 | 3 | 0.0088 | 0.216 |
| *Testing 3* Rotated image with different image | 51 | 2081 | 0 | 0.0088 | 0.2117 |
| *Testing 4* Different image | 52 | 2085 | 0 | 0.0088 | 0.2017 |
| *Testing 5* Different image | 383 | 1997 | 0 | 0.0088 | 0.200 |

SURF but these are linked with higher execution times (4–6 s). The proposed study, employing Blobs + SIFT + ORB + NN, lowers execution time drastically (~ 0.2 s) while ensuring a high key point detection. It provides novel applications such as image authentication, high-resolution data analysis, and digital forensics. This implies that the proposed method is more efficient and flexible for contemporary image analysis tasks.

**Table 6** Comparative result of different algorithm

| Study (year) | Algorithm (s) used | Numbers | | Execution time (second) | Best use case |
|---|---|---|---|---|---|
| | | Blobs | Key points | | |
| Niyishaka and Bhagvati (2018) [9] | DoG + ORB | 239 | 4960 | 4.65 | Image forensics |
| Niyishaka and Bhagvati (2020) [10] | Blobs + Brisk | 239 | 5980 | 6.24 | Image forensics |
| Idris et al. [11] | Blobs + SIFT | 239 | 3027 | 4.10 | Tampered image |
| Yadav and Dongre (2024) [12] | Blobs + SURF | 239 | 3467 | 4.18 | Tampered image |
| *Proposed study* | | | | | |
| Proposed study—same image | Blobs + SIFT + ORB + NN | 21 | 1870 | 0.223 | Image authentication High-resolution data analysis Digital forensics and security |
| Proposed study—rotated image with the same image | | 21 | 1856 | 0.217 | |
| Proposed study—different image | | 52 | 2085 | 0.2017 | |
| Proposed study—rotated image with the different image | | 51 | 2081 | 0.2117 | |

## 6 Conclusion

This study proposes a suitable framework to perform feature extraction and enhancement over large-scale, high-resolution data. Classical approaches, including Harris Corner Detection and Laplacian of Gaussian for the detection procedure, along with Hessian-based segmentation, combine well with key point-based techniques like SIFT and ORB for a better detection scheme and refinement in the descriptors. Experimental results are shown that validate the hybrid approach: feature matching is stable and reliable, even under adverse conditions like rotation, illumination change, and noise. The feature extraction strategy that was presented in this paper, as enhanced as compared to the traditional ones, can achieve greater accuracy and reliability while setting up a scalable basis for applications such as digital forensics, high-resolution data analysis, and image authentication.

## 7 Recommendations

Our research study recommends further exploration of real-world applications, such as deepfake detection for enhancing digital security and medical imaging enhancements for improved diagnostic accuracy. Additionally, future improvements should focus on integrating transformer-based vision models, which have demonstrated superior performance in image analysis tasks. This integration could enhance model efficiency, scalability, and accuracy, making the proposed approach more robust for practical deployment in various domains.

## References


1. Fungtammasan, A., Lee, A., Taroni, J., Wheeler, K., Chin, C., Davis, S., Greene, C.: Ten simple rules for large-scale data processing. PLoS Comput. Biol. **2**, e1009757 (2022). https://doi.org/10.1371/journal.pcbi.1009757
2. Archana, R., Jeevaraj, P.S.E.: Deep learning models for digital image processing: a review. Artif. Intell. Rev. **57**(1) (2024). https://doi.org/10.1007/s10462-023-10631-z
3. Georgiou, T., Liu, Y., Chen, W., Lew, M.: A survey of traditional and deep learning-based feature descriptors for high dimensional data in computer vision. Int. J. Multimed. Inform. Retriev. **9**(3), 135–170 (2019). https://doi.org/10.1007/s13735-019-00183-w
4. Ji, F., Li, G., Lu, S., Ni, J.: Research on a feature enhancement extraction method for underwater targets based on deep autoencoder networks. Appl. Sci. **14**(4), 1341 (2024). https://doi.org/10.3390/app14041341
5. Cheng, G., Lang, C., Wu, M., Xie, X., Yao, X., Han, J.: Feature enhancement network for object detection in optical remote sensing images. J. Rem. Sens. (2021). https://doi.org/10.34133/2021/9805389
6. Li, Z., Dong, Y.: Refined feature enhancement network for object detection. Complex Intell. Syst. **11**(1) (2024). https://doi.org/10.1007/s40747-024-01622-w
7. Huang, L., Yao, C., Zhang, L., Luo, S., Ying, F., Ying, W.: Enhancing computer image recognition with improved image algorithms. Sci. Rep. **14**(1) (2024). https://doi.org/10.1038/s41598-024-64193-3
8. Li, N., Huang, Y., Wang, Z., Fan, Z., Li, X., Xiao, Z.: Enhanced hybrid visio transformer with multi-scale feature integration and patch dropping for facial expression recognition. Sensors **24**(13), 4153 (2024). https://doi.org/10.3390/s24134153
9. Niyishaka, P., Bhagvati, C.: Digital image forensics technique for copy-move forgery detection using DOG and ORB. In: Lecture Notes in Computer Science, pp. 472–483 (2018). https://doi.org/10.1007/978-3-030-00692-1_41
10. Niyishaka, P., Bhagvati, C.: Copy-move forgery detection using image blobs and BRISK feature. Multimed. Tools Appl. **79**(35–36), 26045–26059 (2020). https://doi.org/10.1007/s11042-020-09225-6
11. Idris, B., Abdullah, L.N., Selimun, M.T.A., Halin, A.A.: Copy-move forgery detection using canny edge detector and SIFT-based blob analysis. In: Lecture Notes in Electrical Engineering, pp. 595–614 (2023). https://doi.org/10.1007/978-981-19-8406-8_47
12. Yadav, J., Dongre, N. (eds.): Digital image forgery detection using SURF and ORB technique. Intell. Syst. Appl. Eng. (2024). https://www.ijisae.org/index.php/IJISAE/article/download/4553/3224/9553

# A Study of Generative Artificial Intelligence (GenAI) Capabilities in Generating Supervised Machine Learning Models

**Yousif Abuzuhaira, Mohamed Badawy, and Hasan Kadhem**

**Abstract** Generative AI refers to a subset of artificial intelligence models that generate meaningful content in various forms, such as text, images, and multimedia. The Generative AI industry has witnessed rapid breakthroughs in recent years, with significant attention focused on large language models (LLMs). This paper aims to evaluate and assess the capabilities of Generative AI in generating supervised machine learning models, specifically focusing on four chosen LLMs: OpenAI's ChatGPT, Anthropic's Claude, Google's Gemini, and Microsoft's Copilot. This study employs a multi-phase methodology, where models are trained on various algorithms using traditional methods. Then, the LLMs followed a multi-step prompting strategy, emulating the traditional machine learning pipeline. The LLMs were evaluated based on their responses to direct prompts and their ability to generate Python code. The results show that Generative AI is capable of generating supervised machine learning models, with certain LLMs performing better than others. Claude excelled in preprocessing data, while Gemini performed best in training and testing machine learning models. Overall, there is still room for improvement. The models occasionally panicked and failed to train in a considerable number of instances. These findings could pave the way for further research on the potential contributions of Generative AI to the field of machine learning.

**Keywords** Generative AI · LLM · Machine learning · Supervised ML

Y. Abuzuhaira · M. Badawy (✉) · H. Kadhem
American University of Bahrain, Riffa, Bahrain
e-mail: a01345@aubh.edu.bh

Y. Abuzuhaira
e-mail: a01224@aubh.edu.bh

H. Kadhem
e-mail: Hasan.Kadhem@aubh.edu.bh

 
Y. Ma (ed.), *Proceedings of 4th International Conference on Intelligent Systems Design and Engineering Applications*, Lecture Notes in Electrical Engineering 1478,
https://doi.org/10.1007/978-981-95-1802-9_17

## 1 Introduction

Generative AI is defined as computational techniques that can generate new meaningful content, such as text, images, or multimedia, based on the data they have been trained on [1]. These technologies have proven effective in everyday life, offering substantial assistance across a variety of applications. For instance, OpenAI's ChatGPT serves as an answering engine for daily inquiries, while MidJourney produces artistic outputs, moving beyond mere data mimicry from training datasets. These advancements lay the groundwork for the development of Web 3.0 infrastructure [2]. Generative AI is built on generative models instantiated with machine learning architectures, specifically deep neural networks that create new data samples based on existing patterns. Additionally, the Generative AI system encompasses the entire infrastructure, including the model, data processing, and UI components [1]. The primary objective of this paper is to evaluate the capabilities of Generative AI in generating supervised machine learning models.

Despite extensive research on various subtopics of Generative AI, including its capabilities in programming, image and sound generation, and language proficiency, there is a notable gap in the literature regarding the direct generation of supervised machine learning models. To our knowledge, no prior work has addressed this specific topic. Therefore, this paper aims to explore the milestones that Generative AI has reached in the machine learning field by comparing the capabilities of popular Large Language Models (LLMs) across the supervised machine learning pipeline.

The remainder of the paper is organized as follows. Section 2 discusses the related work on Generative AI for coding and notable contributions to the machine learning field. Section 3 presents information on the two datasets used in the experiment. Section 4 outlines the methodology used for our experiment. Section 5 showcases the results and findings.

## 2 Related Work

Generative AI has made significant impacts in various domains, such as software engineering, literature, graphic design, and many more. The following literature review provides an overview of Generative AI models, their basic purposes and mechanisms, as well as several contributions in the evaluation of large language models in the fields of programming, context understanding, and most importantly, machine learning model creation.

## 2.1 Performance in Programming and Coding

A study titled "Effectiveness of ChatGPT in Coding: A Comparative Analysis of Popular Large Language Models" evaluates the performance of ChatGPT models (GPT 3.5 and GPT 4.5) against other large language models such as Google's Bard and Anthropic's Claude [3]. The research employs a quantitative approach to evaluate the performance of large language models on beginner-level Python concepts. The findings show that ChatGPT significantly outperformed other competitors in multiple programming tasks. While Generative AI can enhance productivity in programming, human oversight is crucial for ensuring accuracy and quality [3].

OpenAI's ChatGPT was tested across 10 programming languages to determine which languages were more suitable for code generation and to identify the overall limitations of Generative AI in code generation [4]. The study focused on metrics such as correctness, efficiency, and maintainability. The results indicated a distribution of effectiveness across various programming languages, with high-level languages like Python and JavaScript performing better than lower-level languages. The performance differences were attributed to the training data available for each language and the syntax of the models. ChatGPT's understanding of prompts was found to be language-dependent, and in some cases, the model failed to understand the context, performing tasks that were explicitly not requested. Additionally, ethical considerations varied by language; for example, a brute force code prompt was deemed unethical in JavaScript, while Ruby produced correct code 9 out of 10 times [4].

In contrast, another study using a pool of LeetCode problems (easy, medium, and hard) as testing prompts for different Generative AI models showed different results [5]. GitHub Copilot outperformed other models by solving 50% of the challenges, excelling in Java and C++ coding problems. Bing AI demonstrated better results in solving Python challenges. Interestingly, ChatGPT was the only model to solve a LeetCode hard problem, although it only solved 20% of the problems correctly. This highlights the need for further refinement of Generative AI models to handle more demanding programming tasks without undermining their significant role in software engineering [1].

## 2.2 Contextual Adaptation and Evolution of Thought in LLMs

The evolution of thought (EoT) is exemplified by the historical belief that the Earth was flat, which was later discovered to be round as philosophy and technology advanced. LLMs have an inherent EoT system that provides contextual results based on the information presented. Each prompt entered by the user is appended to the LLM's history, and the LLM continues to respond based on this context [6, 7]. For instance, if prompted with "Give me a synonym for good," and then queried with

"Bad," the LLM will provide synonyms for "Bad" based on the previous prompt. This demonstrates the importance of providing the LLM with appropriate context to achieve optimal results.

### 2.3 Prompt Engineering Tools

Prompt engineering is a crucial aspect of utilizing Generative AI, as the results can vary drastically depending on the prompt. Tools like MyCrunchGPT and AutoML-GPT have been developed to assist with prompt engineering. MyCrunchGPT is designed for Scientific Machine Learning (SciML), which integrates machine learning with scientific computing to solve engineering problems in fields such as biology, chemistry, and engineering [6]. MyCrunchGPT provides ChatGPT with context and assigns deliverables to produce specific outputs for user prompts. For example, when prompted with "Design 2D airfoil," MyCrunchGPT ensures that the user is prompted for specifications, unlike a generic response from ChatGPT alone [6]. AutoML-GPT is a prototype that streamlines the entire machine learning pipeline, from preprocessing to hyperparameter tuning, utilizing ChatGPT [8]. Users input two cards: a data card containing dataset details and a model card specifying the models and parameters. AutoML-GPT processes these inputs and returns the results, differing from this paper's focus on Generative AI's ability to read and comprehend datasets directly.

### 2.4 LLM Feature Generation

Data scarcity remains a challenge in machine learning projects despite the abundance of data in the digital age. LLMs can mitigate this by generating data for ML algorithms, though the cost depends on the LLM and the volume of data requested. An experiment classified two sets of academic articles using TF-IDF, word embeddings, and features generated by the LLM Llama2 [7]. The LLM-generated features performed comparably to other methods and achieved the best results when combined with TF-IDF features [7]. Another study on identifying good and bad comments in programming code showed that LLM-generated data improved model accuracy [9].

### 2.5 Performance of LLMs

One study [10] measured LLM performance based on their ability to answer questions related to undergraduate knowledge, graduate reasoning, and common knowledge. The LLMs significantly outperformed human benchmarks, with Claude excelling in reasoning and Gemini balancing the two. Another study [11] compared LLM performance in spatial tasks, such as writing code. Although all LLMs lacked code-writing ability, they excelled in task solving, technical skills, and spatial reasoning. ChatGPT, Copilot, and Gemini performed marginally better than Claude in all categories and were more consistent regardless of task complexity. Despite advancements, LLMs have not reached their full potential. They are primarily used for mundane tasks and lack creativity, often mimicking rather than creating new content. LLMs also suffer from "hallucination", generating misleading data when previous patterns do not provide expected results [12]. The quality of LLM output is highly dependent on input quality, which requires users to have experience in prompt engineering to achieve the desired results in complex situations [13]. Running LLMs on a large scale requires significant computing power, which can be costly [13]. Finally, human evaluation remains the most reliable method for validating LLM output, although it is expensive and subject to variance and instability [7].

Based on the literature review, no paper has addressed the ability of machine learning models to generate supervised machine learning models, which is the focus of this paper and its contribution to the fields of machine learning and Generative AI.

# 3 Datasets

## *3.1 Regression Dataset: Laptop Price Prediction*

The regression dataset pertains to predicting laptop prices in Euros [14]. Comprising 1320 rows, this dataset necessitated preprocessing prior to utilization. As detailed in Table 1, it encompasses 9 categorical features and 4 continuous features. Notable

**Table 1** Attributes and data types of the regression dataset

| Attribute | Data type | Description |
|---|---|---|
| Company | String | Laptop manufacturer |
| Product | String | Brand and model |
| TypeName | String | Type (Notebook, Ultrabook, Gaming, etc.) |
| Inches | Numeric | Screen size |
| Screen resolution | String | Screen resolution |
| CPU | String | Central processing Unit (CPU) |
| Ram | String | Laptop RAM |
| Memory | String | Hard disk/SSD Memory |
| GPU | String | Graphics processing units (GPU) |
| OpSys | String | Operating system |
| Weight | String | Laptop weight |
| Price euros | Numeric | Laptop price in euros |

challenges included the operating system column, which contained numerous missing values; the screen resolution column, which exhibited high cardinality due to the inclusion of both display type and resolution; and the CPU column, which also displayed high cardinality. Additionally, the memory column featured variations in units (terabytes and gigabytes) and types (HDD and SSD), while the product and GPU columns exhibited high cardinality.

### 3.2 Classification Dataset: Portuguese Bank Marketing Outcome Prediction

The classification dataset [15], available on Kaggle, comprises real data from a Portuguese bank marketing campaign aimed at promoting bank deposit subscriptions to potential customers. This dataset encompasses 45211 observations, with 39922 (88.3%) classified as 'No' and 5289 (11.7%) classified as 'Yes'. As illustrated in Table 2, it includes a variety of customer demographics and interaction details.

## 4 Methodology

The methodology section outlines the approach taken to evaluate the capabilities of large language models (LLMs) in generating supervised machine learning models. The experiment is divided into two phases. The first phase involves training models on the datasets locally to establish a baseline for comparison. The second phase involves using prompting strategies to evaluate the LLMs' performance.

### 4.1 Phase 1: Training Models Locally

The first phase involved training machine learning models locally on existing datasets using MATLAB R2024B. The machine learning pipeline is illustrated in Fig. 1.

Datasets were imported into the MATLAB environment, and the data handling and cleaning steps were as follows for the two datasets: For the regression dataset:

- Missing values in the operating system column were retained since users could select their own operating system at the time of purchase.
- The screen resolution column was split into resolution and display type. The display type was deleted due to irregular cardinality.
- High cardinality in the CPU column was addressed by shortening entries (e.g., "Intel Core i5 7200U 2.5GHz" was trimmed to "Intel Core i5").

**Table 2** Attributes and data types of the classification dataset

| Attribute | Data type | Description |
|---|---|---|
| Age | Numeric | Client's age |
| Job | Categorical | Type of job (e.g., "admin.", "unemployed", "technician") |
| Marital | Categorical | Marital status ("married", "divorced", "single") |
| Education | Categorical | Level of education ("primary", "secondary", "tertiary", "unknown") |
| Default | Binary | Has credit in default? ("yes", "no") |
| Balance | Numeric | Average yearly balance in euros |
| Housing | Binary | Has a housing loan? ("yes", "no") |
| Loan | Binary | Has a personal loan? ("yes", "no") |
| Contact | Categorical | Contact communication type ("telephone", "cellular", "unknown") |
| Day | Numeric | Last contact day of the month |
| Month | Categorical | Last contact month ("jan" to "dec") |
| Duration | Numeric | Last contact duration in seconds |
| Campaign | Numeric | Number of contacts performed during the campaign (including the last contact) |
| Pdays | Numeric | Days since last contact from a previous campaign (−1 means not previously contacted) |
| Previous | Numeric | Number of contacts before the current campaign |
| Poutcome | Categorical | Outcome of the previous marketing campaign ("unknown", "failure", "success", "other") |
| Y | Binary | Has the client subscribed to a term deposit? ("yes", "no") |

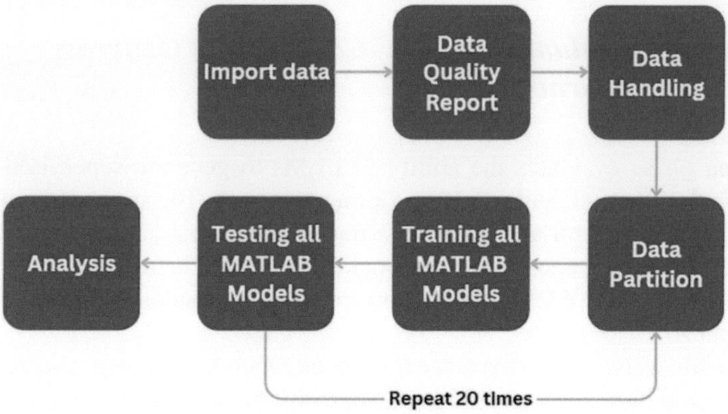


**Fig. 1** Flowchart of phase 1: local machine learning methodology

- The memory column was standardized by converting all instances to gigabytes, and a new binary column was added to indicate whether the memory type is HDD or SSD.
- The product and GPU columns were discarded.

For the classification dataset:

- Undersampling was applied to address the imbalance in the target feature, with only 11.7% of the target feature classified as "yes." All 5289 instances with the "yes" target feature were retained, and a random sample of 14% of the instances with "No" as the target feature was appended.
- The "poutcome" column was deleted due to a large number of missing values.
- The "contact" column was deleted to minimize the degree of freedom and restrict the available features to more relevant ones, such as demographics and transactions.
- Missing values in the "Age" and "education" columns were replaced with the mean value of the respective features.

Prior to model training, the pre-processed datasets were partitioned into two sets, with 70% allocated to the training set and 30% to the testing set. The training sets were trained on all available models from the "Regression Learner" and "Classification Learner" applications in MATLAB. The models trained included Decision Trees, Logistic Regression, Naive Bayes, SVM, Ensemble, Neural Networks, and Kernel of their various presets, totaling over 24 models for classification and 20 models for regression. A 10-fold cross-validation technique was employed to enhance the reliability of the model training. The procedure was repeated 20 times to ensure robustness and consistency in the models and to eliminate sampling bias when drawing conclusions and insights.

### *4.2 Phase 2: Evaluating LLMs' Capability to Generate Machine Learning Models*

The second phase evaluates the ability of LLMs to generate supervised machine learning models through multi-step prompting strategies. The strategies were applied to both the classification and regression datasets to investigate any differences in how the LLMs handle each type. The goal of this phase is to analytically compare the capabilities of the LLMs in data pre-processing, model building, and model performance.

Four LLMs were chosen for this experiment: OpenAI's ChatGPT-4, Anthropic's Claude-3, Google's Gemini-1.5, and Microsoft's Copilot. These models were selected because they are established non-open source models trained on very large datasets by leading companies in the tech industry and were among the earliest language models released as chatbots. Each LLM was tested through a structured series of detailed prompts highlighting the requirements and expected deliverables.

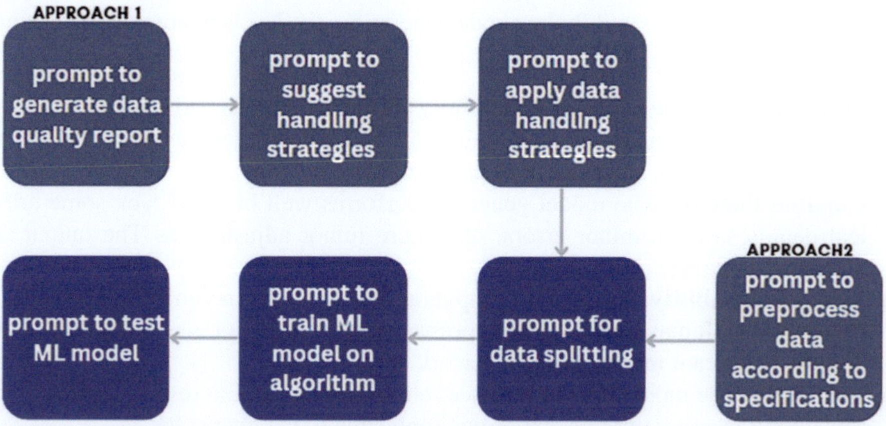


**Fig. 2** Flowchart of phase 2: LLM prompting strategy

The prompts were designed to replicate the machine learning pipeline, mirroring the steps taken in Phase 1, as shown in Fig. 2.

The first prompt instructed the LLM to generate a data quality report, including metrics such as missing values and cardinalities. Another prompt asked the LLM to suggest data handling issues based on the report, followed by a prompt to apply the suggested strategies. An alternative approach was introduced where the LLM was instructed to apply the same preprocessing approaches used in Phase 1. This isolated the data handling and model building abilities from each other. For example, if an LLM failed to identify data handling issues, it would impact the training and testing accuracy. This alternative approach provided another perspective on the LLMs' performance.

After preprocessing, the LLM was asked to partition the dataset into training and testing sets, build a model on the training dataset using a specified algorithm, and then evaluate the trained model on the testing dataset. The experiment was applied to both the classification and regression datasets, resulting in four unique experiments for each LLM. The algorithms were selected based on the highest performing models in Phase 1 and any models recommended by the LLMs. Room was given for model suggestions for improvements, and any viable improvements were taken into account.

**Comparison of Generative AI Models** The comparison is based on the responses generated using the prompting strategies and is conducted across two domains:

- **Direct Prompts**: This measures the degree to which each model can produce direct instructions or task-specific output (e.g., data quality reports, model training summaries) based on human-readable prompts.
- **Code Generation in Python**: This assesses the model's ability to generate Python code that is functionally and syntactically correct for a given task, and evaluates the correctness, efficiency, and best practices of the generated code.

For both direct prompts and Python code generation, the levels of comparison are as follows:

- **Fully Capable**: The model produces fully correct, complete, and working outputs or code without visible errors or omissions. The content generated is relevant to the task and reflects a high level of understanding.
- **Capable Partials**: The model generally performs well but may lack some critical details, contain minor errors, or require minor adjustments. The output is understandable and generally useful.
- **Limited Capability**: The model's outputs are substantially incomplete, inaccurate, or incoherent. It may not address the primary task elements or generate results that require significant manual fixes or rework.
- **Incapable**: The model fails to produce relevant or applicable results for the task. The output is either not relevant or not applicable to the given task.

## 5 Results and Discussion

Two devices were utilized for this experiment, with each allocated to one dataset. For the regression dataset, an ROG Zephyrus G15 AMD Ryzen 9 5900HS with Radeon Graphics 3.30 GHz and 16GB RAM was used. For the classification dataset, a Lenovo IdeaPad Gaming 3 with 16.0 GB RAM and an AMD Ryzen 7 7735HS with Radeon Graphics 3.20 GHz was employed.

### 5.1 *Phase 1: Training Models Locally*

**Regression Dataset** For the regression dataset, the best average accuracy across 20 iterations was achieved by Rational Quadratic Gaussian Process Regression (GPR), with an accuracy of 81.74%. GPR demonstrated the highest consistency in predicting laptop prices, maintaining readings equal to or above 80% in most iterations, with rare instances of 79%. Quadratic SVM also exhibited consistent accuracy, reaching 86% in one iteration. The least accurate model was Medium Tree, which occasionally achieved 80% accuracy but mostly hovered around 73%. The accuracies of all algorithms are detailed in Table 3.

**Classification Dataset** For the classification dataset, the majority of the models achieved accuracy above the 80% mark. Table 4 presents the highest-performing models and their average accuracy across 20 iterations. Bagged Trees and Medium Gaussian SVM were the top performers, with average accuracies of 83.89% and 83.84%, respectively.

The distribution of the F1 score closely mirrors the distribution of the accuracy, as shown in the boxplots in Fig. 3. Both boxplots are very similar, indicating that

**Table 3** Performance metrics of regression models

| Module | Accuracy (%) |
|---|---|
| Rational quadratic GPR | 81.74 |
| Matern 5/2 GPR | 81.71 |
| Exponential GPR | 81.70 |
| Quadratic SVM | 81.20 |
| Squared exponential GPR | 81.06 |
| Bagged trees | 77.77 |
| Boosted trees | 77.47 |
| Medium gaussian SVM | 76.71 |
| Linear | 75.58 |
| Medium tree | 73.62 |

**Table 4** Performance metrics of classification models

| Model | Accuracy (%) |
|---|---|
| Bagged trees | 83.89 |
| Medium gaussian SVM | 83.84 |
| Quadratic SVM | 83.24 |
| Narrow neural network | 83.21 |
| Boosted trees | 83.02 |
| Bilayered neural network | 82.47 |
| Trilayered neural network | 82.44 |
| SVM kernel | 82.31 |
| Logistic regression kernel | 82.19 |
| Cubic SVM | 81.98 |
| Fine tree | 81.75 |
| Coarse gaussian SVM | 81.22 |
| Medium neural network | 80.88 |
| Linear SVM | 80.69 |
| RUSBoosted trees | 80.63 |
| Binary GLM logistic regression | 80.25 |
| Medium tree | 80.25 |

the models' performance in terms of accuracy and F1 score is consistent. Medium Gaussian SVM and Bagged Trees achieved the highest recorded accuracy and F1 score, with Medium Gaussian SVM reaching a maximum of 84.95% for both metrics. No model exceeded the 90% mark, which contrasts with the results reported in a previous study using the same dataset [16]. This discrepancy might be attributed to the undersampling applied to the dataset prior to training.

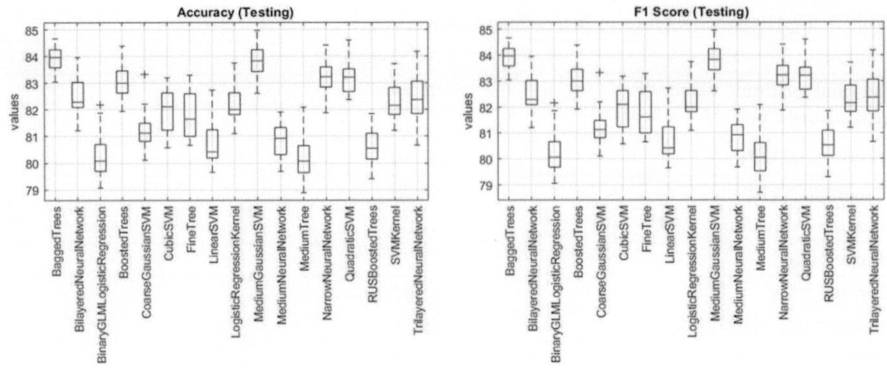


**Fig. 3** Boxplots for highest performing models for classification dataset

## *5.2 Phase 2: Evaluation of LLMs*

**Data Quality Report Generation** When reading the CSV data, the simple "read_csv" function did not work as intended unless a keyword argument 'encoding=latin-1' or 'encoding=ISO-8859-1' was added. ChatGPT and Gemini both detected the error in the code and regenerated code that allowed the LLM to read the data sufficiently. In contrast, Claude and Copilot did not adjust the code as intended, requiring manual changes to the read line.

The data quality report needed to be separated into numerical and categorical features. The LLMs approached this by detecting the data type from the CSV file. If the data type was int64 or float64, it was treated as continuous; otherwise, it was treated as categorical. This approach led to issues, such as the "weight" feature in the regression dataset, which had rows inputted as "2.2kg." Although it is numerically obvious, its data type in the CSV file was object, so it was treated as categorical. Additionally, when finding missing values, the LLMs utilized the isnull() function, which returns whether the argument is null or not. This caused issues with features like "poutcome" in the classification dataset, which had rows containing values named "unknown." While "unknown" logically means missing, the LLMs did not comprehend this because they used the isnull() function and treated it as a present value in the dataset.

**Data Quality Issues and Handling Strategies** The strategies that ChatGPT, Gemini, and Copilot used to address data quality issues were very static and did not demonstrate a deep understanding of basic preprocessing principles. For instance, these LLMs often replaced missing values with the mean or median without considering the percentage of missing data. If 80% of the data were missing, the LLMs would not remove the entire column; instead, they would replace the missing values with the mode or mean, as shown in Fig. 4.

Furthermore, these three LLMs used the same approach to handle outliers in both continuous and categorical features. For continuous features, the LLMs capped

```
# Analyze numerical features
for col in numerical_df.index:
    # Missing Values Check
    missing_percentage = numerical_df.loc[col, 'Missing Percentage']
    if missing_percentage > 0:
        issues_and_strategies.append({
            'Feature Name': col,
            'Issue': f"{missing_percentage:.2f}% missing values",
            'Handling Strategy': "Impute missing values with median or mean"
        })
```

**Fig. 4** Example of LLM-generated preprocessing code

outliers at a certain percentile if the row was 3 standard deviations away from the mean. Similarly, for categorical features, outliers were placed in a "rare" category or "other" category. In contrast, ClaudeAI excelled in this area because it did not apply the same logic to every column. Claude detected issues that the other LLMs could not find, such as the need to split certain columns. Overall, the first three LLMs took a very similar static approach to handling data quality issues, while Claude showed a deeper comprehension of advanced levels of data preprocessing.

**Applying Handling Strategies** When it came to applying handling strategies, some LLMs were not able to apply the strategies they proposed. For example, for some features in the regression dataset, Copilot proposed handling strategies such as "ensuring that the rows do not include any bias" or "investigating further if this is expected or a data entry error." However, when prompted to apply these strategies, Copilot returned the same statements, as shown in Tables 5 and 6. Similar issues were observed with ChatGPT and Gemini when dealing with the imbalance in the classification dataset. Despite detecting the imbalance, these LLMs did not meet expectations when prompted to apply the proposed strategies. Overall, the LLMs performed their handling strategies adequately.

**Cleaning Regression and Classification Datasets with Our Instructions** When asked to clean the datasets using our instructions, the LLMs performed very well and mostly produced the exact output that was manually produced. The main issue was with the regression dataset, where the LLMs were instructed to split the Memory column into memory type and memory size while standardizing the units of the memory size. Although the LLMs split the column, they did not succeed in unifying the units of the new memory size column. Additionally, Copilot could not handle the classification dataset because it was too large, meaning it could not be tested on how Copilot would handle the classification dataset. Overall, the LLMs performed well and produced the desired output.

**Splitting Data into Training and Testing** The LLMs had no issues splitting the data into training and testing sets. For the classification dataset, the LLMs split the data so that the ratio of the target feature was the same for both training and testing.

**Table 5** Handling strategies by copilot on the regression dataset

| Feature name | Issue | Handling strategy |
|---|---|---|
| Company | Second mode frequency is equal to mode frequency | Investigate further to understand if this is expected or a data entry error |
| Cpu | High cardinality with 118 unique values | Consider grouping similar CPUs into broader categories |
| Gpu | High cardinality with 110 unique values | Consider grouping similar GPUs into broader categories |
| Memory | High cardinality with 39 unique values | Consider grouping similar memory sizes into broader categories |
| OpSys | High frequency of 'Windows 10' (82.27%) | Ensure that the dominance of 'Windows 10' does not bias the model |
| Product | High cardinality with 618 unique values | Consider removing this feature if it does not add significant value to the model |
| Ram | High frequency of '8GB' (47.51%) | Ensure that the dominance of '8GB' does not bias the model |
| ScreenResolution | High cardinality with 40 unique values | Consider grouping similar screen resolutions into broader categories |
| TypeName | High frequency of 'Notebook' (55.7%) | Ensure that the dominance of 'Notebook' does not bias the model |
| Weight | High cardinality with 179 unique values | Consider grouping similar weights into broader categories |
| Inches | Outliers detected beyond 3 standard deviations | Remove outliers or cap them at 3 standard deviations from the mean |

However, for the regression dataset, the LLMs did not take a consistent approach. Sometimes they split the data so that the mean, median, and standard deviation were approximately the same, and other times they performed a stratified split to ensure that the target feature was split into consistent bins for training and testing, which were then reverted back to normal continuous data.

**Results** Based on the entries in Tables 7 and 8 , ChatGPT performed fairly consistently but mainly had issues handling data quality issues and failed to consider feature generation. Furthermore, ChatGPT struggled with applying hyperparameter tuning. Copilot significantly underperformed due to its token limitations, which made it unable to read the classification dataset, and it overall produced poor results in training and testing. Claude AI excelled in detecting and addressing data quality issues, being the only LLM in this paper to consider feature generation and column splitting. However, it faced challenges in training and testing ML models. Gemini

**Table 6** Applied data handling strategies by copilot on the regression dataset

| Feature name | Applied strategy |
|---|---|
| Company | Investigate further to understand if this is expected or a data entry error |
| Cpu | Grouped similar CPUs into broader categories |
| Gpu | Grouped similar GPUs into broader categories |
| Memory | Grouped similar memory sizes into broader categories |
| OpSys | Ensure that the dominance of 'Windows 10' does not bias the model |
| Product | Removed feature due to high cardinality |
| Ram | Ensure that the dominance of '8GB' RAM does not bias the model |
| ScreenResolution | Grouped similar screen resolutions into broader categories |
| TypeName | Ensure that similar 'Notebook' does not bias the model |
| Weight | Grouped similar weights into broader categories |
| Inches | Removed outliers beyond 3 standard deviations |

performed the best in training and testing, based on its results having scored high accuracies overall. Although it performed well in training and testing, it took a linear approach to handling data quality issues and did not manage to mention feature imbalances in the classification dataset, similar to ChatGPT and Copilot.

Tables 9 and 10 cover the regression results for the LLM preprocessing and instructed preprocessing, respectively, while Tables 11 and 12 display the classification results for the LLM preprocessing and instructed preprocessing. ChatGPT performed best in the regression dataset using Random Forest with its own preprocessing (84% accuracy) and worst with Gradient Boosting using instructed preprocessing (73% accuracy). It did not create models for all algorithms but produced results similar to MATLAB. In classification, it performed best with Random Forest and instructed preprocessing (86% accuracy, 86% F1 score) and worst with SVM and its own preprocessing (83% accuracy, 54% F1 score). Copilot performed best in regression using Random Forest with its own preprocessing (78% accuracy) but failed to create models for some algorithms and could not generate any models for instructed preprocessing or classification. Furthermore, Copilot was not able to read the classification dataset due to token limitations, so it provided no results. ClaudeAI excelled in regression using Random Forest with its own preprocessing (93% accuracy), however, performed poorly with Gradient Boosting utilizing instructed preprocessing (50% accuracy). It was effective at preprocessing but did not produce models for all algorithms. In classification, it performed best with Random Forest and instructed preprocessing (85% accuracy, 86% F1 score) and worst with SVM and its own preprocessing (88% accuracy, 40% F1 score). Gemini was the most consistent, achieving 90% accuracy in regression with Random Forest and its own preprocessing, while its worst result was 75% with Linear Regression and instructed preprocessing. It performed well in classification, similar to ClaudeAI, and was the only LLM that nearly created models for all algorithms. However, like the others, it achieved high accuracy due to imbalanced data in some cases.

**Table 7** LLMs performance in direct prompts

| Direct prompts | Regression | | | | Classification | | | |
|---|---|---|---|---|---|---|---|---|
| Task | ChatGPT | Copilot | Claude | Gemini | ChatGPT | Copilot | Claude | Gemini |
| Data quality report generation | Partially | Partially | Partially | Partially | Partially | Incapable | Partially | Partially |
| Data quality issues and handling strategy | Limited | Limited | Fully | Limited | Limited | Incapable | Fully | Limited |
| Apply data quality handling strategy | Fully | Limited | Fully | Fully | Partially | Incapable | Fully | Partially |
| Data cleaning instructions | Partially | Partially | Partially | Partially | Fully | Incapable | Fully | Fully |
| Dataset split (80% training, 20% testing) | Fully | Fully | Fully | Fully | Fully | Incapable | Fully | Fully |
| Train a machine learning model | Limited | Limited | Limited | Partially | Partially | Incapable | Partially | Partially |
| Test a machine learning model | Limited | Limited | Limited | Partially | Partially | Incapable | Partially | Partially |

**Table 8** LLMs performance in the python code generation

| Generating python code | Regression | | | | Classification | | | |
|---|---|---|---|---|---|---|---|---|
| Task | ChatGPT | Copilot | Claude | Gemini | ChatGPT | Copilot | Claude | Gemini |
| Data quality report generation | Fully | Fully | Fully | Fully | Fully | Incapable | Fully | Fully |
| Data quality issues and handling strategy | Fully | Fully | Fully | Fully | Fully | Incapable | Fully | Fully |
| Apply data quality handling strategy | Fully | Limited | Fully | Fully | Fully | Incapable | Fully | Fully |
| Data cleaning instructions | Fully | Fully | Fully | Fully | Fully | Incapable | Fully | Fully |
| Dataset split (80% training, 20% testing) | Fully | Fully | Fully | Fully | Fully | Incapable | Fully | Fully |
| Train a machine learning model | Limited | Limited | Limited | Partially | Partially | Incapable | Partially | Partially |
| Test a machine learning model | Limited | Limited | Limited | Partially | Partially | Incapable | Partially | Partially |

**Table 9** Accuracy metrics for the regression dataset that was self preprocessed by the LLM, NC = Not Capable

| Regression dataset preprocess | ChatGPT | | Copilot | | Claude | | Gemini | |
|---|---|---|---|---|---|---|---|---|
| Algorithm | Train | Test | Train | Test | Train | Test | Train | Test |
| SVM | NC | NC | NC | NC | NC | NC | 80% | 80% |
| Random forest (%) | 80 | 84 | 73 | 78 | 83 | 93 | 80 | 79 |
| KNN | NC | NC | NC | NC | NC | NC | NC | NC |
| Gradient Boosting | 79% | 82% | NC | NC | NC | NC | NC | NC |
| Linear Regression | NC | NC | 75% | 70% | 68% | NC | 74% | 75% |

**Table 10** Accuracy metrics for the regression dataset that was preprocessed according to our instructions, NC = Not Capable

| Regression dataset instructed preprocess | ChatGPT | | Copilot | | Claude | | Gemini | |
|---|---|---|---|---|---|---|---|---|
| Algorithm | Train | Test | Train | Test | Train | Test | Train | Test |
| SVM | NC | NC | NC | NC | NC | NC | 78% | 85% |
| Random Forest | 77% | 78% | NC | NC | 65% | 50% | 91% | 90% |
| KNN | NC | NC | NC | NC | NC | NC | NC | NC |
| Gradient Boosting | 78% | 73% | NC | NC | NC | NC | NC | NC |
| Linear Regression | NC | 75% | NC | NC | 48% | 50% | 86% | NC |

**Table 11** Accuracy and F1 scores metrics for the classification dataset that was self preprocessed by the LLM

| Classification dataset preprocess | GPT train | | GPT test | | Claude train | | Claude test | | Gemini train | | Gemini test | |
|---|---|---|---|---|---|---|---|---|---|---|---|---|
| Algorithm | Accuracy | F1 | Accuracy | F1 | Accuracy | F1 | Accuracy | F1 | Accuracy | F1 | Accuracy | F1 |
| Random forest (%) | 90 | 48 | 90 | 49 | 92 | 49 | 91 | 43 | 92 | 49 | 91 | 43 |
| SVM (%) | 83 | 55 | 83 | 54 | 89 | 42 | 88 | 40 | 89 | 42 | 88 | 40 |
| Logistic regression (%) | 84 | 55 | 84 | 54 | 90 | 44 | 89 | 43 | 90 | 44 | 89 | 43 |

**Table 12** Accuracy and F1 scores metrics for the classification dataset that was preprocessed according to out instructions

| Classification dataset instructed preprocess | GPT train | | GPT test | | Claude train | | Claude test | | Gemini train | | Gemini test | |
|---|---|---|---|---|---|---|---|---|---|---|---|---|
| Algorithm | Accuracy | F1 | Accuracy | F1 | Accuracy | F1 | Accuracy | F1 | Accuracy | F1 | Accuracy | F1 |
| Random forest (%) | 83 | 85 | 86 | 86 | 84 | 85 | 85 | 86 | 84 | 85 | 85 | 85 |
| SVM (%) | 73 | 71 | 73 | 71 | 83 | 84 | 84 | 85 | 72 | 70 | 72 | 70 |
| Logistic regression (%) | 80 | 82 | 80 | 80 | 80 | 80 | 81 | 81 | 80 | 80 | 81 | 81 |

## References

1. Feuerriegel, S., Hartmann, J., Janiesch, C., Zschech, P.: Generative ai. Business and Inf. Syst. Eng. **66**(1), 111–126 (2023). https://link.springer.com/article/10.1007/s12599-023-00834-7
2. Bandi, A., Adapa, P.V.S.R., Kuchi, Y.E.V.P.K.: The power of generative ai: a review of requirements, models, input–output formats, evaluation metrics, and challenges. Future Internet **15**(8). https://www.mdpi.com/1999-5903/15/8/260
3. Coello, C.E.A., Alimam, M.N., Kouatly, R.: Effectiveness of chatgpt in coding: a comparative analysis of popular large language models. Digital **4**(1), 114–125 (2024). https://www.mdpi.com/2673-6470/4/1/5
4. Buscemi, A.: A comparative study of code generation using chatgpt 3.5 across 10 programming languages (2023). https://arxiv.org/abs/2308.04477
5. Idrisov, B., Schlippe, T.: Program code generation with generative ais. Algorithms **17**(2) (2024), https://www.mdpi.com/1999-4893/17/2/62
6. Kumar, V., Gleyzer, L., Kahana, A., Shukla, K., Karniadakis, G.E.: Mycrunchgpt: A llm assisted framework for scientific machine learning. J. Mach. Learn. Model. Comput. **4**(4), 41–72 (2023)
7. Balek, V., SÃžkora, L., Sklenák, V., Kliegr, T.: Llm-based feature generation from text for interpretable machine learning (2024), https://arxiv.org/abs/2409.07132
8. Zhang, S., Gong, C., Wu, L., Liu, X., Zhou, M.: Automl-gpt: automatic machine learning with gpt (2023). https://arxiv.org/abs/2305.02499
9. Akl, H.A.: A ml-llm pairing for better code comment classification (2023). https://arxiv.org/abs/2310.10275
10. Akpan, M.: Have we reached agi? comparing chatgpt, claude, and gemini to human literacy and education benchmarks. https://arxiv.org/abs/2407.09573
11. Hochmair, H., Juhász, L., Kemp, T.: Correctness comparison of chatgpt-4, gemini, claude-3, and copilot for spatial tasks. Trans. GIS **28**(7), 2219–2231 (2024). https://onlinelibrary.wiley.com/doi/abs/10.1111/tgis.13233
12. Morris, C., Jurado, M., Zutty, J.: Llm guided evolution—the automation of models advancing models. In: Proceedings of the Genetic and evolutionary computation conference. GECCO '24, Association for Computing Machinery, New York, NY, USA, pp 377–384. https://doi.org/10.1145/3638529.3654178
13. Zhang, X., Zhang, J., Rekabdar, B., Zhou, Y., Wang, P., Liu, K.: Dynamic and adaptive feature generation with llm. https://arxiv.org/abs/2406.03505
14. Varli, M.: Laptop price (2020). https://www.kaggle.com/datasets/muhammetvarl/laptop-price, retrieved from Kaggle, https://www.kaggle.com/datasets/muhammetvarl/laptop-price
15. Patra, S.: Bank marketing campaign (2021). https://www.kaggle.com/datasets/edith2021/bank-marketing-campaign?select=bank.csv, data retrieved from Kaggle, https://www.kaggle.com/datasets/edith2021/bank-marketing-campaign?select=bank.csv
16. Moro, S., Cortez, P., Laureano, R.: Using data mining for bank direct marketing: an application of the crisp-dm methodology. In: Proceedings of the European simulation and modelling conference

MIX
Papier aus verantwortungsvollen Quellen
Paper from responsible sources
FSC® C105338

If you have any concerns about our products,
you can contact us on
**ProductSafety@springernature.com**

In case Publisher is established outside the EU,
the EU authorized representative is:
**Springer Nature Customer Service Center GmbH**
**Europaplatz 3, 69115 Heidelberg, Germany**

Printed by Libri Plureos GmbH
in Hamburg, Germany